The Science of Fractal Images

Heinz-Otto Peitgen   Dietmar Saupe
Editors

The Science of Fractal Images

Michael F. Barnsley   Robert L. Devaney   Benoit B. Mandelbrot
Heinz-Otto Peitgen   Dietmar Saupe   Richard F. Voss

With Contributions by
Yuval Fisher   Michael McGuire

With 142 Illustrations in 277 Parts and 39 Color Plates

Springer-Verlag New York Berlin Heidelberg
London Paris Tokyo

Heinz-Otto Peitgen
Institut für Dynamische Systeme, Universität Bremen, D-2800 Bremen 33, Federal Republic
of Germany, *and*
Department of Mathematics, University of California, Santa Cruz, CA 95064, USA

Dietmar Saupe
Institut für Dynamische Systeme, Universität Bremen, D-2800 Bremen 33, Federal Republic
of Germany

The cover picture shows a fractal combining the two main topics of this book : deterministic and random fractals. The deterministic part is given in the form of the potential surface in a neighborhood of the Mandelbrot set. The sky is generated using random fractals. The image is produced by H. Jürgens, H.-O. Peitgen and D. Saupe.

The back cover images are: Black Forest in Winter (top left, M. Barnsley, F. Jacquin, A. Malassenet, A. Sloan, L. Reuter); Distance estimates at boundary of Mandelbrot set (top right, H. Jürgens, H.-O. Peitgen, D. Saupe); Floating island of *Gulliver's Travels* (center left, R. Voss); Foggy fractally cratered landscape (center right, R. Voss); Fractal landscaping (bottom, B. Mandelbrot, K. Musgrave).

Library of Congress Cataloging-in-Publication Data
The Science of fractal images : edited by Heinz-Otto Peitgen
and Dietmar Saupe ; contributions by Michael F. Barnsley . . . [et al.].
   p. cm.
  Based on notes for the course Fractals—introduction, basics, and
perspectives given by Michael F. Barnsley, and others, as part of
the SIGGRAPH '87 (Anaheim, Calif.) course program.
  Bibliography: p.
  Includes index.
  1. Fractals. I. Peitgen, Heinz-Otto, 1945– . II. Saupe, Dietmar,
1954– . III. Barnsley, M. F. (Michael Fielding), 1946–
QA614.86.S35 1988
516—dc19   88–12683

This book was prepared with LaTeX on Macintosh computers and reproduced by Springer-Verlag from camera-ready copy supplied by the editors.
TEX is a trademark of the American Mathematical Society. Macintosh is a trademark of Apple Computer, Inc.
Printed and bound by Arcata Graphics/Halliday, West Hanover, Massachusetts.
Printed in the United States of America.

9 8 7 6 5 4 3 2

ISBN 0-387-96608-0 Springer-Verlag New York Berlin Heidelberg
ISBN 3-540-96608-0 Springer-Verlag Berlin Heidelberg New York

Preface

This book is based on notes for the course *Fractals:Introduction, Basics and Perspectives* given by Michael F. Barnsley, Robert L. Devaney, Heinz-Otto Peitgen, Dietmar Saupe and Richard F. Voss. The course was chaired by Heinz-Otto Peitgen and was part of the SIGGRAPH '87 (Anaheim, California) course program. Though the five chapters of this book have emerged from those courses we have tried to make this book a coherent and uniformly styled presentation as much as possible. It is the first book which discusses fractals solely from the point of view of computer graphics. Though fundamental concepts and algorithms are not introduced and discussed in mathematical rigor we have made a serious attempt to justify and motivate wherever it appeared to be desirable. Basic algorithms are typically presented in pseudo-code or a description so close to code that a reader who is familiar with elementary computer graphics should find no problem to get started.

Mandelbrot's fractal geometry provides both a description and a mathematical model for many of the seemingly complex forms and patterns in nature and the sciences. Fractals have blossomed enormously in the past few years and have helped reconnect pure mathematics research with both natural sciences and computing. Computer graphics has played an essential role both in its development and rapidly growing popularity. Conversely, fractal geometry now plays an important role in the rendering, modelling and animation of natural phenomena and fantastic shapes in computer graphics.

We are proud and grateful that Benoit B. Mandelbrot agreed to write a detailed foreword for our book. In these beautiful notes the *Father of Fractals* shares with us some of the computer graphical history of fractals.

The five chapters of our book cover :
- an introduction to the basic axioms of fractals and their applications in the natural sciences,
- a survey of random fractals together with many pseudo codes for selected algorithms,
- an introduction into fantastic fractals, such as the Mandelbrot set, Julia sets and various chaotic attractors, together with a detailed discussion of algorithms,
- fractal modelling of real world objects.

V

Chapters 1 and 2 are devoted to random fractals. While Chapter 1 also gives an introduction to the basic concepts and the scientific potential of fractals, Chapter 2 is essentially devoted to algorithms and their mathematical background. Chapters 3, 4 and 5 deal with deterministic fractals and develop a dynamical systems point of view. The first part of Chapter 3 serves as an introduction to Chapters 4 and 5, and also describes some links to the recent chaos theory.

The Appendix of our book has four parts. In Appendix A Benoit B. Mandelbrot contributes some of his brand new ideas to create random fractals which are directed towards the simulation of landscapes, including mountains and rivers. In Appendix B we present a collection of magnificent photographs created and introduced by Michael Mc Guire, who works in the tradition of Ansel Adams. The other two appendices were added at the last minute. In Appendix C Dietmar Saupe provides a short introduction to rewriting systems, which are used for the modelling of branching patterns of plants and the drawing of classic fractal curves. These are topics which are otherwise not covered in this book but certainly have their place in the computer graphics of fractals. The final Appendix D by Yuval Fisher from Cornell University shares with us the fundamentals of a new algorithm for the Mandelbrot set which is very efficient and therefore has potential to become popular for PC based experiments.

Almost throughout the book we provide selected pseudo codes for the most fundamental algorithms being developed and discussed, some of them for beginning and some others for advanced readers. These codes are intended to illustrate the methods and to help with a first implementation, therefore they are not optimized for speed.

The center of the book displays 39 color plates which exemplify the potential of the algorithms discussed in the book. They are referred to in the text as *Plate* followed by a single number N. Color plate captions are found on the pages immediately preceding and following the color work. There we also describe the front and back cover images of the book. All black and white figures are listed as *Figure N.M*. Here N refers to the chapter number and M is a running number within the chapter.

After our first publication in the *Scientific American*, August 1985, the Mandelbrot set has become one of the brightest stars of amateur mathematics. Since then we have received numerous mailings from enthusiasts around the world.

We have reproduced some of the most beautiful experiments (using *MSetDEM( )*, see Chapter 4) on pages 20 and 306. These were suggested by David Brooks and Daniel N. Kalikow, Framingham, Massachusetts.

Bremen, March 1988

Heinz-Otto Peitgen and Dietmar Saupe

Acknowledgements

Michael F. Barnsley acknowledges collaboration with Laurie Reuter and Alan D. Sloan, both from Georgia Institute of Technology. Robert L. Devaney thanks Chris Frazier from the University of Bremen for putting his black and white figures in final form. Chris Frazier also produced Figures 0.1, 5.1, 5.2 and 5.17. Heinz-Otto Peitgen acknowledges collaboration with Hartmut Jürgens and Dietmar Saupe, University of Bremen, and thanks Chris Frazier for some of the black and white images. Dietmar Saupe thanks Richard F. Voss for sharing his expertise on random fractals. Besides the authors several people contributed to the color plates of this book, which is gratefully acknowledged. A detailed list of credits is given in the Color Plate Section.

Michael F. Barnsley
Robert L. Devaney
Yuval Fisher
Benoit B. Mandelbrot
Michael McGuire
Heinz-Otto Peitgen
Dietmar Saupe
Richard F. Voss

Michael F. Barnsley. *1946 in Folkestone (England). Ph. D. in Theoretical Chemistry, University of Wisconsin (Madison) 1972. Professor of Mathematics, Georgia Institute of Technology, Atlanta, since 1983. Formerly at University of Wisconsin (Madison), University of Bradford (England), Centre d'Etudes Nucleaires de Saclay (Paris). Founding officer of Iterated Systems, Inc.

Robert L. Devaney. *1948 in Lawrence, Mass. (USA). Ph. D. at the University of California, Berkeley, 1973. Professor of Mathematics, Boston University 1980. Formerly at Northwestern University and Tufts University. Research interests: complex dynamics, Hamiltonian systems.

M.F. Barnsley   *R.L. Devaney*

Yuval Fisher. *1962 in Israel. 1984 B.S. in Mathematics and Physics, University of California, Irvine. 1986 M.S. in Computer Science, Cornell University. 1988 Ph. D. in Mathematics (expected), Cornell University.

Benoit B. Mandelbrot. *1924 in Warsaw (Poland). Moved to Paris in 1936, to USA in 1958. Diploma 1947, Ecole Polytechnique, D. Sc. 1952, Paris, Dr. Sc. (h. c.) Syracuse, Laurentian, Boston, SUNY. 1974 I.B.M. Fellow at Thomas J. Watson Research Center and 1987 Abraham Robinson Adjunct Professor of Mathematical Science, Yale University. Barnard Medal 1985, Franklin Medal 1986. Member of the American Academy of Arts and Sciences and of the U.S. National Academy of Sciences.

Y. Fisher   *B.B. Mandelbrot*

Michael McGuire. *1945 in Ballarat, Victoria (Australia). Ph. D. in Physics, University of Washington (Seattle), 1974. He has worked in the field of atomic frequency standards at the University of Mainz, Germany, NASA, Goddard Space Flight Center, and Hewlett Packard Laboratories.

Heinz-Otto Peitgen. *1945 in Bruch (Germany). Dr. rer. nat. 1973, Habilitation 1976, both from the University of Bonn. Research on nonlinear analysis and dynamical systems. 1977 Professor of Mathematics at the University of Bremen and since 1985 also Professor of Mathematics at the University of California at Santa Cruz. Visiting Professor in Belgium, Italy, Mexico and USA.

M. McGuire   *H.-O. Peitgen*

Dietmar Saupe. *1954 in Bremen (Germany). Dr. rer. nat. 1982 at the University of Bremen. Visiting Assistant Professor of Mathematics at the University of California, Santa Cruz, 1985–87 and since 1987 at the University of Bremen. There he is a researcher at the Dynamical Systems Graphics Laboratory. Research interests : mathematical computer graphics and experimental mathematics.

Richard F. Voss. *1948 in St. Paul, Minnesota (USA). 1970 B. S. in Physics from M.I.T. 1975 Ph. D. in Physics from U. C. Berkeley. 1975–present: Research Staff Member at the I.B.M. Thomas J. Watson Research Laboratory in Yorktown Heights, NY. Research in condensed matter physics.

D. Saupe   *R.F. Voss*

VIII

Contents

The future   Springer-Verlag logo?

Foreword

People and events behind the "Science of Fractal Images"

Benoit B.Mandelbrot

It is a delight to watch Heinz-Otto Peitgen get together with several of our mutual friends, and tell the world the secrets of drawing fractals on the computer. A pedant would of course proclaim that the very first publication in each branch of fractals had immediately revealed *every* secret that matters. So let me rephrase what was said: this book's goal is to tell the world how to draw the basic fractals without painfully rediscovering what is already known.

The book needs no foreword, but being asked to provide one, without limitation of space, has unleashed a flood of recollections about some Men and some Ideas involved in the *Science of Fractal Images,* including both Art for Art's sake and Art for the sake of Science. A few of these recollections may even qualify as history, or perhaps only as what the French call *la petite histoire.* As some readers may already know, for me history is forever part of the present.

Perhaps as a reward for holding this belief, the very fact of writing down for this book my recollections concerning fractal forgery of landscapes has made me actually unhappy again about a feature of all past fractal forgeries, that they fail to combine relief with rivers. Eventually, we did something about this defect, as well as about other features of the subdivision forgeries described in the body of this book. The new directions are sketched in Appendix A and were added to this book at the last minute.

0.1   The prehistory of some fractals-to-be: Poincaré, Fricke, Klein and Escher

To begin, while fractal geometry dates from 1975, it is important in many ways to know that a number of shapes now counted as fractals have been known for a long time. But surprisingly few had actually been drawn before the computer era. Most were self-similar or self-affine and represent the artless work of the draftsmen on the payroll of science publishers. Also, there are renditions of physical and simulated Brownian motion in the book by Jean Perrin, *Les Atomes*, and William Feller's *Introduction to Probability*. These renditions have helped me dream in fruitful ways (as told in my 1982 book *The Fractal Geometry of Nature* [68] p. 240), but they are not beautiful. Fractals-to-be occur in the work of Fatou and Julia circa 1918, but they led to no illustration in their time.

However, Poincaré's even earlier works circa 1890 do include many sketches, and two very different nice stories are linked with illustrations that appeared shortly afterwards, in the classic book titled *Vorlesungen über die Theorie der automorphen Funktionen* [43], which Fricke & Klein published in 1897. This book's text and its preface are by the hand of Fricke, R. Robert Fricke, but (see p. vi) the great Felix Klein, "a teacher and dear friend" seems to have graciously consented to having his name added on the title page. The illustrations became even more famous than the text. They have been endlessly reproduced in books on mathematics, and for the better or for the worse have affected the intuition of countless mathematicians.

A tenacious legend claims that students in industrial drawing at the Technische Hochschule in Braunschweig, where Fricke was teaching mathematics, drew these figures as assignment, or perhaps even as an exam. Unkind words have been written about some of the results. In fact, I have done my share in detailing the defects of those which claim to represent the fractal-to-be limit sets of certain Kleinian groups (leading some to wonder which of Fricke's students should be failed posthumously). These dubious figures were drawn with the help of the original algorithm of Poincaré, which is very slow, too slow even for the computer. However, my paper [70] in *The Mathematical Intelligencer* in 1983 has given an explicit and quick new algorithm for constructing such limit sets, as the complements of certain " sigma-discs", and has compared Fricke's Figure 156 with the actual shape drawn by a computer program using the new algorithm. The comparison is summarized in *The Fractal Geometry of Nature*, page 179. As was to be expected, the actual shape is by far the more detailed and refined of the two, but this is not all: against all expectations, it is *not* nec-

Fig. 0.1: Circle Limits IV by M.C. Escher, ©1988 M.C. Escher c/o Cordon Art – Baarn – Holland

essarily perceived as being more complicated. I feel it is more harmonious, and can be comprehended as a whole, therefore it is perceived as far simpler than the clumsy old pictures. However, a famous mathematician (15 years my senior) has expressed dismay at seeing the still vibrant foundation of his intuition knocked down by a mere machine.

Of wider popular interest by far are Fricke's drawings of "hyperbolic tessellations", the reason being that they have become widely popular behind diverse embellishments due to the pen of Maurits C. Escher, As seen for example in the book *The World of M.C. Escher* [33]. Many people immediately perceive some "obvious but hard to describe" connection between Escher and fractals, and it is good to know that these tessellations are indeed closely related to fractals. In fact, they were knowingly triggered by Poincaré, as is well documented by H.S.M. Coxeter in his *Leonardo* [22] paper of 1979. Having seen some of Escher's early work, this well-known geometer wrote to him and received the following answer: "Did I ever thank you ...? I was so pleased with this booklet and proud of the two reproductions of my plane patterns!... Though the text of your article [in *Trans. Royal Soc. Canada*, 1957] is much too learned for a simple, self-made plane pattern-man like me, some to the illustrations... gave me quite a shock. ... Since a long time I am interested in patterns with "motives" getting smaller and smaller till they reach the limit of infinite smallness... but I was never able to make a pattern in which each "blot" is getting smaller gradually from a center towards the outside circle-limit, as [you] show.... I tried to find out how this figure was geometrically constructed, but I succeeded only in finding the centers and radii of the largest inner-circles. If you could give me a simple explanation..., I should be immensely pleased and very thankful to you! Are there other systems besides this one to reach a circle-limit? Nevertheless,... I used your model for a large woodcut". This was his picture 'Circle Limit I', concerning which he wrote on another occasion: "This woodcut *Circle Limit I*, being a first attempt, displays all sorts of shortcomings".

In his reply, Coxeter told Escher of the infinitely many patterns which tessellated a Euclidean or non-Euclidean plane by black and white triangles. Escher's sketch-books show that he diligently pursued these ideas before completing *Circle Limits II, III, IV*. He wrote: "In the coloured woodcut *Circle Limit III* most of the defects [of *Circle Limit I*], have been eliminated". In his *Magic Mirror of M.C. Escher* (1976), Bruno Ernst wrote: "best of the four is *Circle Limit III*, dated 1959... In addition to arcs placed at right angles to the circumference (as they ought to be), there are also some arcs that are not so placed". [Now going back to Coxeter], "In fact all the white arcs 'ought' to cut the circumference at the same angle, namely 80° (which they do, with remarkable accuracy). Thus Escher's work, based on his intuition, without any computation, is perfect, even though his poetic description of it was only approximate".

The reader is encouraged to read Coxeter's paper beyond these brief quotes. But an important lesson remains. As already stated, the Coxeter pictures which made Escher adopt the style for which he became famous, hence eventually affected the esthetics of many of our contemporaries, were not the pure creation of an artist's mind. They came straight from Fricke & Klein, they were largely inspired by Henri Poincaré, and they belong to the same geometric universe as fractals. Note also that the preceding story is one of only two in this paper to involve a person who had been professionally trained as an artist.

0.2   Fractals at IBM

The first steps of the development of a systematic fractal geometry, including its graphic aspects, were taken at the IBM T.J. Watson Research Center, or wherever I happened to be visiting from this IBM base. The next task, therefore, in historical sequence, is to reminisce about the IBM fractals project.

This project has always been an example of very small science, in fact it had reduced to myself for the first ten of my thirty years at IBM. Since then, it has in principle included one full-time programmer; actually, there were short overlaps and long periods with no programmer. The assistants of J.M. Berger (whom I had "borrowed' in 1962), as well as my project's first assistant, Hirsh Lewitan, were "career" IBM employees, but all the others were recent graduates or even students on short contract. Here is a complete chronological list of those who stayed for over a few weeks: G.B.Lichtenberger (part-time), M.S.Taqqu, J.L.Oneto, S.W.Handelman, M.R.Laff, P.Moldave (part-time), D.M.McKenna, J.A.Given, E.Hironaka, L.Seiter, F.Guder, R.Gagné and K. Musgrave. The grant of IBM Fellowship in 1974 also brought a half-time secretary: H.C.Dietrich, then J.T. Riznychok, and later V.Singh, and today L.Vasta is my full-time secretary.

R.F.Voss has been since 1975 an invaluable friend, and (as I shall tell momentarily) a close companion when he was not busy with his low-temperature physics. The mathematicians J.Peyriére, J.Hawkes and V.A.Norton, and the meteorologist S. Lovejoy (intermittently) have been post-doctoral visitors for a year or two each, and two "IBM'ers", the hydrologist J.R.Wallis and the linguist F.J.Damerau, have spent short periods as de facto inter-project visitors. As for equipment, beyond slide projectors, terminals and P.C.'s and (of course) a good but not lavish allotment of computer cycles, my project has owned one high-quality film recorder since 1983. Naturally, a few IBM colleagues outside of my project have also on occasion briefly worked on fractals.

These very short lists are worth detailing, because of inquiries that started coming in early in 1986, when it was asserted in print, with no intent to praise, that "IBM has spent on fractals a perceptible proportion of its whole research budget". The alumni of the project are surprised, but endlessly proud, that the bizarre perception that fractals ever became big science at IBM should be so widely accepted in good faith. But the almost threadbare truth is even more interesting to many observers of today's scientific scene. To accept it, and to find it deserving gratitude, was the price paid for academic freedom from academia.

The shortness of these lists spanning twenty years of the thirty since I joined IBM also explains my boundless gratitude for those few people.

0.3   The fractal mountains by R.F. Voss

My next and very pleasant task is to tell how I met the co-authors of this book, and some other people who matter for the story of the *Science of Fractal Images*.

During the spring of 1975, Richard F. Voss was hopping across the USA in search of the right job. He was soon to become Dr. Voss, on the basis of a Berkeley dissertation whose contents ranged from electronics to music, without ever having to leave the study of a widespread physical phenomenon (totally baffling then, and almost equally baffling today), called $\frac{1}{f}$-*noise*. Other aspects of this noise, all involving fractals, were favorites of mine since 1963, and my book *Les objets fractals,* which was to be issued in June 1975, was to contain primitive fractal mountains based on a generalization of $\frac{1}{f}$-noise from curves to surfaces. One of the more striking parts of Voss's thesis concerned (composed) music, which he discovered had many facets involving $\frac{1}{f}$-noises. He had even based a micro-cantata on the historical record of Nile river discharges, a topic dear to my heart.

Therefore, Voss and I spoke after his job-hunting talk at IBM Yorktown, and I offered a deal: come here and let us play together; something really nice is bound to come out. He did join the Yorktown low-temperature group and we soon became close co-workers and friends. Contrary to what is widely taken for granted, he never joined my tiny project, and he has spent the bulk of his time on experimental physics. Nevertheless, his contribution to fractals came at a critical juncture, and it has been absolutely essential. First, we talked about writing a book on $\frac{1}{f}$-noise, but this project never took off (and no one else has carried it out, to my knowledge). Indeed, each time he dropped by to work together, he found me involved with something very different, namely, translating and revising *Les objets fractals.* The end result came out in 1977 as *Fractals,* and preparing it kept raising graphics problems. Voss ceaselessly inquired about what Sig Handelman and I were doing, and kept asking whether we would consider better ways, and then he found a sure way of obtaining our full attention.

He conjured a computer graphics system where none was supposed to exist, and brought along pictures of fractals that were way above what we had been dealing with until then. They appeared in *Fractals,* which is why the foreword describes him as the co-author of the pictures in that book.

Color came late at Yorktown, where it seems we fractalists continued to be the only ones to use demanding graphics in our work. We first used color in my next book, the 1982 *Fractal Geometry of Nature.* In late 1981, the text was already in the press, but the color pictures had not yet been delivered to the publishers. The film recorder we were using was ours on a short lease, and this fact and everything else was conspiring to make us rush, but I fought back. Since "the desire is boundless and the act a slave to limit" ([68], p. 38), I fought hardest for the sake of the *Fractal Planetrise* on the book's jacket. It was soon refined to what (by the standards of 1981) was perfection, but this was not enough. Just another day's work, or another week's, I pleaded, and we shall achieve something that would not need any further improvement, that would not have to be touched up again when "graphic lo-fi" will go away, to be replaced by "graphic hi-fi". To our delight, this fitted Voss just fine.

Fractal illustrations had started as wholly unitarian, the perceived beauty of the old ones by Jean-Louis Oneto and Sig Handelman being an unexpected and unearned bonus. But by 1981 their beauty had matured and it deserved respect, even from us hard scientists, and it deserved a gift of our time. Many people have, since those days, showed me their fractal pictures by the hundreds, but I would have been happier in most cases with fewer carefully worked out ones.

Everyone experiences wonder at Voss's pictures, and "to see [them] is to believe [in fractal geometry]". Specialists also wonder how these pictures were done, because, without ever drawing specific attention to the fact, Voss has repeatedly conjured technical tricks that were equivalent to computer graphics procedures that did not officially develop until much later. This brings to mind a philosophical remark.

Watching Voss the computer artist and Voss the physicist at work for many years had kept reminding me of the need for a fruitful stress between the social and the private aspects of being a scientist. The only civilized way of being a scientist is to engage in *the process* of doing science primarily for one's *private pleasure* . To derive pleasure from the *public results* of this process is a much more common and entirely different matter. The well-known danger is that, while *dilettare* means *to delight* in Italian, its derivative *dilettante* is a term of contempt. While not a few individually profess to be serious scientists, yet motivated primarily by personal enjoyment of their work, very few could provide what I view as the only acceptable evidence of "serious dilettantism". This is a willingness and perhaps even a compulsion to leave significant portions of one's best work unpublished or unheralded — knowing full well that one could claim no credit for them. This may be easiest for the scientific giants; Lars Onsager was a legend on this account. On the other hand, every scientist has been the active or the passive witness of episodes when one could not or would not work in a field without becoming thoroughly uncivilized. The true test, therefore, arises when civilized behavior is neither easy not impossible. On these and other stringent grounds, I view Dick Voss (as graphics expert and as physicist) as being one of the most civilized serious scientists in my wide acquaintance.

0.4   Old films

What about films? We were ill-equipped to produce them, having only an exhausted film recorder (a tube-based Stormberg-Carlson 4020) at our disposal. In 1972, however, with Hirsh Lewitan, we did prepare a clip on the creation of fractal galaxy clusters, using the *Seeded Universe* method. Then, in 1975, with Sig Handelman, we added a clip in which the landscape to be later used on Plate 271 of *The Fractal Geometry of Nature* emerged slowly from the deep, then rotated majestically (or at least very slowly), and finally slipped back under water. Spontaneously, everyone called this the *Flood* sequence. By a fluke, the highest altitude was achieved on two distinct points, and a programming flaw stopped the Flood when these points were still visible. Delighted, I indulged in commenting that my fractal model of relief had predicted that there were two tips to Mount Ararat, not one, ... until an auditor straight from Armenia reported very drily that this fact was well-known to everyone in his country. The *Galaxy Clustering* and the *Mount Ararat Mount Ararat* sequences, taken together, were nicknamed *Competing with the Good Lord on Sunday.* They soon came to look out-of-date and pitifully primitive, but now they are of historical interest,... valuable antiques.

The *Flood,* and a recent animation of one of Voss's data bases, done by R.Greenberg Associates, both moved around a landscape, without zooming.

0.5   Star Trek II

But what about the "real Hollywood" ? "It" realized immediately the potential of Voss's landscape illustrations in my 1977 book, and soon introduced variants of these fractals into its films. This brought a lovely reenactment of the old

and always new story of the Beauty and the Beast, since it is taken for granted
that films are less about Brains than about Beauty, and since the few brainy
mathematicians who had known about individual fractals-to-be had taken for
granted (until my books) that these were but Monsters, ... Beastly. The pillars
of "our geographically extended Hollywood" were Alain Fournier, Don Fussell
and Loren Carpenter. Early in 1980, John W. van Ness, a co-author of mine in
1968 who had moved to the University of Texas at Dallas, asked me to comment
on the draft of his student Fournier's Ph.D. dissertation. Fournier and Fussell
had written earlier to us asking for the IBM programs to generate fractal mountains, but we did not want to deal with lawyers for the sake of programs that were
not documented, and were too intimately linked to one set of computers to be
readily transported anywhere else. Therefore, Fournier and Fussell went their
own way, and soon hit upon an alternative method that promised computations
drastically faster than those of Voss.

Precisely the same alternative was hit upon at the same time by Loren Carpenter, then at Boeing Aircraft, soon afterwards to move to Lucasfilm, and now
at Pixar. In his own words in *The College Mathematics Journal* for March 1984,

"I went out and bought The Fractal Geometry of Nature *as soon as
I read Martin Gardner's original column on the subject in* Scientific
American. *I have gone through it with a magnifying glass two or
three times. I found that it was inspirational more than anything
else. What I got out of it myself was the notion that "Hey, these
things are all over, and if I can find a reasonable mathematical
model for making pictures, I can make pictures of all the things
fractals are found in. That is why I was quite excited about it. . . .*

*"The method I use is recursive subdivision, and it has a lot of ad-
vantages for the applications that we are dealing with here; that is,
extreme perspective, dynamic motion, local control — if I want to
put a house over here, I can do it. The subdivision process involves
a recursive breaking-up of large triangles into smaller triangles.
We can adjust the fineness of the precision that we use. For example, in "Star Trek", the images were not computed to as fine
a resolution as possible because it is an animated sequence and
things are going by quickly. You can see little triangles if you look
carefully, but most people never saw them.*

*"Mandelbrot and others who have studied these sorts of processes
mathematically have long been aware that there are recursive approximations to them, but the idea of actually using recursive ap-*

proximations to make pictures, a computer graphics-type applica-
tion, as far as we know first occurred to myself and Fournier and
Fussel, in 1979....

"One of the major problems with fractals in synthetic imagery is the
control problem. They tend to get out of hand. They will go random
all over the place on you. If you want to keep a good tight fist on it
and make it look like what you want it to look like, it requires quite
a bit of tinkering and experience to get it right. There are not many
people around who know how to do it."

While still at Boeing, Carpenter became famous in computer graphics circles for making a short fractal film, *Vol Libre,* and he was called to Lucasfilm to
take a leading role in the preparation of the full feature *Star Trek II: The Wrath of
Khan.* Several computer-generated sequences of this film involve fractal landscapes, and have also become classics in the core computer graphics community.
The best known is the *Genesis* planet transformation sequence. A different com-
pany, Digital Productions, later included massive fractal landscapes in *The Last
Starfighter,* which I saw — without hearing it — in an airplane. I had seen *Star
Trek* in a suburban movie-house (since I had gone there on duty, my stub was
reimbursed). An associate had seen it on a previous day, and had reported that
it was too bad that the fractal parts have been cut (adding as consolation that is
was known that they *always* cut out the best parts in the suburbs). Of course, my
wife and I immediately saw where the fractal portion started, and we marveled:
If someone less durably immersed than the two of us in these matters could be
fooled so easily, what about people at large?

Later, interviewed for the summer 1985 issue of *La lettre de l'image,* Carpenter described the severe cost constraints imposed by his work: "We cannot
afford to spend twice as much money to improve the quality of the pictures
by 2%". We would hate to be asked to attach a numerical % to quality improvement, but computer costs do keep decreasing precipitously, and we hope
that future feature films using fractals will be cheap while pleasing even to the
crankiest among mathematicians.

This Beauty and the Beast episode was most enjoyable, but also drew us into
a few scrapes, long emptied of bitterness, but instructive. We were disappointed
that the endless credits of the films never included the word *fractal,* nor our
names. The excuse was that everyone who mattered knew, so there was no
need to say anything. Besides, lawyers feared that, if mentioned, we would
have been put in a position to sue for a part of the cake. The world at large
does not believe that scientists are resigned to the fact that their best work —

the principles of mathematics and the laws of nature — cannot be patented, copyrighted, or otherwise protected. All that the scientists can expect is to be paid in the coin of public — not private — praise.

Later on, we greeted with amusement Alvy Ray Smith's term "graftal". The differences from "fractal" were hardly sufficient to justify this proprietary variation on my coinage.

Fournier, Fussel and Carpenter are not represented in this book. It is a pity that we did not come to know them better. They have hardly ever written to us, even at a time when we could have helped, and would have loved to do so, and anyhow would have liked to follow their work as it evolved.

0.6 Midpoint displacement in Greek geometry: The Archimedes construction for the parabola

Our scrapes with "our Hollywood" have led to a variety of mutually contradictous impressions. Some people came to believe that the fractal landscapes of Fournier, Fussel and Carpenter are, somehow, not "true fractals". Of course *they are fractals,* just as true as the Koch curve itself. Other people believe that I begrudge credit for "recursive subdivision" in order to claim "midpoint displacement" — which is the same thing under a different term — for myself. Actually, as the French used to be taught in high school geometry, the basic credit for the procedure itself (but of course not for fractals) belongs to someone well beyond personal ambition, namely to Archimedes (287–212 BC). The antiquity of the reference is a source of amusement and wonder, but rest assured that his work is amply documented. One of his great achievements was to evaluate the area between a parabola and a chord AB, and many writers view his argument as the first documented step towards calculus. Write the parabola's equation as $y = P(x) = a - bx^2$, with directions chosen so that $b > 0$. Given the chord's endpoints $\{\, x_A, P(x_A) = a - bx_A^2 \,\}$ and $\{\, x_B, P(x_B) = a - bx_B^2 \,\}$, Archimedes interpolates $P(x)$ recursively to values of x that form an increasingly tight dyadic grid. In a first step, observe that

$$
\begin{aligned}
P\left[\tfrac{1}{2}(x_A + x_B)\right] \;-\; & \tfrac{1}{2}[\,P(x_A) + P(x_B)\,] = \\
= \;& a - \tfrac{b}{4}(x_A + x_B)^2 - \left[a - \tfrac{b}{2}(x_A^2 + x_B^2)\right] \\
= \;& \tfrac{b}{4}(x_B - x_A)^2 \\
= \;& \tfrac{\delta}{4}
\end{aligned}
$$

(by definition of δ). Thus, the first stage of interpolation requires an upward displacement of $\frac{\delta}{4}$ to be applied to the midpoint of the original chord, replacing it by two shorter ones. In the second stage of interpolation, the counterpart of $x_B - x_A$ is twice smaller, hence this stage requires an upward displacement of the midpoint of each sub-chord, by the amount equal to $4^{-2}\delta$. Etc. ... The k-th stage requires an upward displacement of the midpoints of 2^{k-1} chords by the amount equal to $4^{-k}\delta$. Of course, the idea that the parabola has an equation was not known until Descartes devised analytic geometry. However, an ingenious ad-hoc argument had allowed Archimedes to derive the above rule of upward displacements equal to $4^{-k}\delta$.

0.7   Fractal clouds

The algorithm Voss used to generate fractal mountains extends to clouds, as described in his contribution to this book. The resulting graphics are stunning, but happen not to provide an adequate fit to the real clouds in the sky. This is the conclusion we had to draw from the work of Shaun Lovejoy.

Lovejoy, then a meteorology student in the Physics Department at McGill University in Montreal, wrote to me, enclosing a huge draft of his thesis. The first half, concerned with radar observation, was not controversial and sufficed to fulfill all the requirements. But the second half, devoted to the task of injecting fractals in meteorology, was being subjected to very rough weather by some referees, and he was looking for help. My feeling was that this work showed very great promise, but needed time to "ripen". (I was reminded of my own Ph.D. thesis, which had been hurried to completion in 1952; I was in a rush to take a post-doctoral position, a reason that soon ceased to appear compelling). Hence, my recommendation to Lovejoy was that he should first obtain his sheepskin on the basis of his non-controversial work, then work with me as a post-doctoral student. We argued that he must not leave in his publications too many points that the unavoidable unfriendly critics could latch on to.

I liked best Shaun's area-perimeter diagram, drawn according to fractal precepts in my 1977 book, which suggested that the perimeters of the vertical projections of clouds (as seen from zenith, for example from a satellite) are of fractal dimension about $\frac{4}{3}$. A paper reduced to this diagram and a detailed caption appeared in *Science* in 1982, and immediately became famous. A second of many parts of Lovejoy's thesis required far more work, and finally I pitched in. The clouds in our joint paper, which came out (years later) in *Tellus* for 1985 (see [62]), do not seem to have yet been surpassed. By then, Lovejoy had drifted away from us. He had grown impatient with my refusal to reopen old fights that had been won to an acceptable degree, and by my deliberate preference for seeking "soft acceptance", with controversy only when it is unavoidable, as opposed to "hard acceptance", with unforgiving victims.

To landscape painters, clouds seem to pose a severe challenge, but one has achieved fame for his prowess. His name was Salomon van Ruysdael (1602-1670), and he brings to mind a question and a story. The question is whether fractal geometry can help compare the clouds by Ruysdael and those by Mother Nature. The story does not *really* belong here, because it does not involve the computer as artist's tool, but let me go ahead. Elizabeth Carter was an undergraduate in meteorology at the University of California at Los Angeles (UCLA), in the group of Professor George L. Siscoe. Her hobby is photographing clouds, and she had found a nice way of getting academic credit for it. They evaluated the fractal dimension for many varied clouds' contours (as seen from a nearly horizontal direction, which is not the same thing as Lovejoy's views from the zenith). They found that Nature's clouds' D varies over an unexpectedly broad range, while Ruysdael's clouds' D is far more tightly bunched. In hindsight, the result was as expected: the painter chose to paint clouds that are dramatic, yet not impossible, hence his clouds' D's are near Nature's maximum.

0.8 Fractal trees

Before moving into non linear fractals, it seemed logical to me as manager of a tiny fractals group, to perform a few preliminary tests without perturbing the on-going programs. This is how a Princeton senior, Peter Oppenheimer, came to work with us for a few weeks. Later he wrote his senior thesis on fractals, and eventually he moved to the New York Institute of Technology on Long Island, and became an expert on fractal botany. Today he encounters competition from Przemyslaw Prusinkiewicz.

Drawing non-random fractal trees is comparatively easy, and there are several in *The Fractal Geometry of Nature*. But drawing random fractal trees that are not of unrealistic "sparseness" presents a major difficulty, because branches must not overlap. Suppose that a random tree is to be constructed recursively. One cannot add a branch, or even the tiniest twig, without considering the Euclidean neighborhood where the additions will be attached. However, points that are close by according to Euclidean distance may be far away according to the graph distance taken along the branches. Therefore, a random recursive construction of a tree, going from trunk to branches and on to twigs, is by necessity a global process. One may be drawn to seeking a construction by self-contemplation, or by obeying the constraints imposed by one's computer better way.

By contrast, space appears forgiving, more precisely offers a near-irresistible temptation to cheat. Indeed, show a shape described as a tree's projection on a plane, and challenge our mind to imagine a spatial tree having such a projection. Even when the original spatial branches happen to intersect or to become entangled, our mind will readily disentangle them, and see them as a tree.

Now back to planar trees, and to ways of drawing them without worrying about self-intersection. A completely natural method was devised by Tom Witten and Leonard Sander. It came about in what we think is the best possible way, not during a search for special effects, but during a search for scientific understanding of certain web or tree-like natural fractal aggregates. The Witten-Sander method is called *diffusion limited aggregation.* Most unfortunately, it fails to yield realistic botanical trees, but it gives us hope for the future.

0.9 Iteration, yesterday's dry mathematics and today's weird and wonderful new fractal shapes, and the "Geometry Supercomputer Project"

Now, from fractals that imitate the mountains, the clouds and the trees, let us move on to fractals that do not. For the artist and the layman, they are simply weird and wonderful new shapes. My brief involvement with Poincaré limit sets has already been touched upon. My initiation to Julia sets began at age 20, when the few who knew them called them J-sets. This episode, and the beginning of my actual involvement with the study of the iteration of $z \to z^2 + c$, have both been described in an Invited Contribution to *The Beauty of Fractals* [83] which need not be repeated here.

But I do want to mention in this Foreword a brief interaction with David Mumford, which eventually contributed to a very interesting and broad recent development.

David's name was known to me, as to everybody else in mathematics, because of his work in algebraic geometry. We met when I came to Harvard in 1979, and in November 1979 he came to a seminar I gave. After the talk, which was on iteration, he rushed towards me : – On the basis of what you have said, you should also look into Kleinian groups; you might even find an explicit construction for their limit set. – Actually, I responded, I already have a nice algorithm for an important special case. Please come to my office, and I shall show you. He came over and saw the algorithm that was eventually published in *The Mathematical Intelligencer* in 1983, as told earlier in this Foreword. – This is so simple, that Poincaré should have seen it, or someone else since Poincaré. Why did the discovery have to wait for you? – Because no one before me had used a powerful new tool, the computer! – But one cannot prove anything with a computer! – Sure, but playing with the computer is a source of conjectures, often most unexpected ones. The conjecture it has suggested about Kleinian limit sets has been easy to prove; others are too hard for me. – In that case, would you help me to learn to play with the computer? – With pleasure, but we would have to get help from my latest IBM assistant, Mark Laff.

Soon after, it became clear that Mumford had to seek associates closer by, in Cambridge. He was tutored by my course assistant Peter Moldave, and started working with David Wright, then a Harvard graduate student in mathematics, who ceased at that point to hide his exceptional programming skills. Eventually, Mumford became thoroughly immersed in computers, first as heuristic tools, then for their own sake.

He became instrumental in helping the awareness of the computer-as-tool spread among mathematicians. The resulting needs grew so rapidly that, after a lapse of hardly eight years, the National Science Foundation has now established a *Geometry Supercomputer Project*! The charter members are F. Almgren (Princeton), J. Cannon (Brigham Young), D. Dobkin (Princeton), A. Douady (ENS, Paris), D. Epstein (Warwick), J. Hubbard (Cornell), B.B. Mandelbrot (IBM and Yale), A. Marden (Minnesota), J. Milnor (IAS, Princeton), D. Mumford (Harvard), R. Tarjan (Princeton and Bell Labs), and W. Thurston (Princeton). At the risk of sounding corny, let me confess that the opening of this project was a high point in my life.

The next topic to be discussed concerning iteration is my very fruitful interaction with V. Alan Norton, a Princeton mathematics Ph.D. in 1976, who was in my group as a post-doc in 1980-82, and stayed on the research staff at Yorktown. He was one of the two principal "illustrators" of *The Fractal Geometry of Nature,* as seen in that book's very detailed picture credits. He has achieved great renown, starting with SIGGRAPH 1982 for his splendid quaternionic Julia set pictures.

Norton also worked on the end-papers of *The Fractal Geometry of Nature,* on which hangs a tale that may be worth recounting. These end-papers (which the book's early printings leave without legend, an omission I now regret) involve an important problem from the theory of iteration of analytic functions, an artifact due to inherent limitations of the computer, and two decorative touches.

Fig. 0.2: The Julia set of Newton's method applied to $e^z = 1$.

The original graph was unbounded, and Norton introduced a decorative touch: to invert with respect to a circle. I loved the result; unfortunately, while bounded, it did not fit neatly on a double page spread. Hence I imposed a second and more arbitrary decorative touch: to stretch the graph horizontally to fill the space available.

The serious mathematical problem that had led me to this graph was the use of Newton's method to solve the equation $exp(z) = c$. The solutions are known from calculus, but Gaston Julia has shown in 1917 that Newton's method is a fruitful ground to study the iteration of functions of a complex variable z. Chapter 19 of *The Fractal Geometry of Nature* examines the iteration of $z^2 + c$ and other polynomials. This end-paper relates to the iteration of the transcendental function $z - 1 + ce^{-z}$ (see also Figure 0.2).

In Arthur Cayley's pioneering *global* studies of iteration, in 1879, the interest in iteration had arisen from the application of the Newton method. (Peitgen et al. tell the story, and illustrate it, in *The Mathematical Intelligencer* in 1984[84].) Cayley began by solving $z^2 = c$, which proved easy, then went on to try $z^3 = c$, which stumped him by exhibiting three "grey areas" that he found no way of resolving. Julia in 1917 had found many facts about these areas, and John H. Hubbard had shown us his revealing earliest graph of the corresponding Julia set. It was natural for us, in late 1980, to play with $z^p = c$, and then view $e^z = c$ as a suitable limit of $z^p = c$ for $p \to \infty$. We made many interesting observations on this limit case, but the study was far from completed and publishable when we moved on to very different work.

Final and unfortunate fact, the non-fractal bold boundaries between the background and the solidly colored areas on the end-papers of *The Fractal Geometry of Nature* are an artifact. The study of transcendental functions' iterates leads very quickly to enormous integers, hence soon reaches intrinsic limits beyond which the computer takes its own arbitrary actions.

0.10 Devaney, Barnsley and the Bremen "Beauty of Fractals"

Our perception of the iteration of transcendental functions as a difficult and very rich topic was confirmed by several eminent mathematicians, such as Robert L. Devaney. No wonder, therefore, that one should see striking resemblances between our end-papers and his beautiful and widely seen illustrations and films. Bob's papers on the iteration of transcendental functions had already brought admirative attention to him, but we did not become fast friends until we started bumping into each other constantly on the fractals *Son et Lumière* traveling shows.

The life orbit of Michael Barnsley has also crossed mine, and then stayed in the same neighborhood, because of fractals. The amusing background, in this instance, is in the public record, and I should not repeat it. I first read about it in James Gleick's book, *Chaos: The Birth of a New Science.* It told me how it came to be that Michael burst into my house one day, full of enthusiasm, and of lovely tales. Later, we held a few meetings at the Atlanta airport (of all places!), and since then it has been a pleasure to keep up with his work and that of his many associates.

Now back to the pictures of Julia and Mandelbrot sets in *The Fractal Geometry of Nature.* During the summer of 1984, we were tooling up to redo them in color, with Eriko Hironaka as programmer, when mail brought in, hot off the press, the June issue of the German magazine *Geo.* We realized immediately that much of what we were proposing had already been achieved, in fact achieved beyond our aspirations, by Heinz-Otto Peitgen, Peter H. Richter, Dietmar Saupe and their associates. This group's earlier fractal pictures in *The Mathematical Intelligencer* earlier in 1984 had been most welcome, but those in color had not yet given reason for enthusiasm. The color pictures in the 1984 *Geo* showed a skilled and artistic eye, and a sure hand, one that had gained experience but had not become a lazy or a hasty one. They were unmistakably the outcome of the search for perfection I had admired earlier in the work of Voss, and always attempt in my own.

I wrote to the *Geo* authors at the University of Bremen to congratulate them, to tell them of the change of plans their success had provoked, and to express the hope of meeting them soon. They told me about a fractal exhibit they were planning, and for which they were preparing a catalogue that eventually led to their book *The Beauty of Fractals,* and they asked me to write the personal contribution mentioned earlier in this Foreword. Granted that this book was fated not to be done by us, it is a delight that it came from them. I gained these new friends when they invited me to Bremen in May 1985, to open the first showing of their exhibit, and I participated again in this exhibit in several other cities. Our joint appearances since then have become too numerous to count. There are no anecdotes to tell, only very pleasant events to remember.

Conclusion

As my friends team up for this book under Heinz-Otto's and Dietmar's editorship, I am reminded that only a while ago (the hurt is thoroughly gone, but the memory remains) no one wanted to scan my early pictures for longer than a few minutes, and this would-be leader of a new trend had not a single follower. Then (very slowly in old memory, yet almost overnight in present perception) fractals had become so widely interesting that SIGGRAPH started devoting lectures, then full day courses to them. The first makeshift fractals at SIGGRAPH came in 1985 under my direction, the second were in 1986 under Peter Oppenheimer, and the third in 1987 have led to the present volume.

Is this book likely to be the last word on the subject? I think not. Several outstanding authors were left out because of the book's origin. And my own experience, that the act of writing foreword has sufficed to make me aware of many new open directions to explore, is bound to be repeated very widely. Let us all pay to the book the high compliment of promptly making it quite obsolete.

Chapter 1

Fractals in nature: From characterization to simulation

Richard F. Voss

Mandelbrot's fractal geometry provides both a description and a mathematical model for many of the seemingly complex forms found in nature. Shapes such as coastlines, mountains and clouds are not easily described by traditional Euclidean geometry. Nevertheless, they often possess a remarkable simplifying invariance under changes of magnification. This statistical self-similarity is the essential quality of fractals in nature. It may be quantified by a fractal dimension, a number that agrees with our intuitive notion of dimension but need not be an integer. In Section 1.1 computer generated images are used to build visual intuition for fractal (as opposed to Euclidean) shapes by emphasizing the importance of self-similarity and introducing the concept of fractal dimension. These fractal forgeries also suggest the strong connection of fractals to natural shapes. Section 1.2 provides a brief summary of the usage of fractals in the natural sciences. Section 1.3 presents a more formal mathematical characterization with fractional Brownian motion as a prototype. The distinction between self-similarity and self-affinity will be reviewed. Finally, Section 1.4 will discuss independent cuts, Fourier filtering, midpoint displacement, successive random additions, and the Weierstrass-Mandelbrot random function as specific generating algorithms for random fractals. Many of the mathematical details and a discussion of the various methods and difficulties of estimating fractal dimensions are left to the concluding Section 1.6.

1.1   Visual introduction to fractals: Coastlines, mountains and clouds

The essence of fractals is illustrated in Figure 1.1 with successive views of a fractal planet from an orbiting spacecraft. In the search for a suitable landing site, a portion of the initial view as shown in the upper left of Figure 1.1 is magnified by a factor of 4 to give the middle top image. Similarly, each successive image represents a magnification of a selected portion of the coastline of the previous image (as indicated by the white box) up to a final magnification of more than 10^7 in the middle bottom. This high magnification can be compared with the initial image that is repeated at the lower right. Although these two images, differing in scale by more than 10^7, are not identical, they seem to share so many of the same characteristics that it is difficult to believe that they are not just different sections of the same landscape at the same magnification. This property of objects whereby magnified subsets look like (or identical to) the whole and to each other is known as *self-similarity*. It is characteristic of fractals and differentiates them from the more traditional Euclidean shapes (which, in general, become ever smoother upon magnification).

Figure 1.2 shows another computer generated forgery. Its three elementary fractal shapes, the foreground landscape, its distribution of craters, and the generalization of Brownian motion onto a sphere rising in the background, all share this characteristic self-similarity. Magnified subsets of each of these shapes are again reminiscent of the whole. The fact that these computer generated primitives of fractal geometry strongly evoke the real world, is a valuable clue to their importance in Nature.

According to Galileo (1623),

> *Philosophy is written in this grand book - I mean universe - which stands continuously open to our gaze, but it cannot be understood unless one first learns to comprehend the language in which it is written. It is written in the language of mathematics, and its characters are triangles, circles and other geometrical figures, without which it is humanly impossible to understand a single word of it; without these, one is wandering about in a dark labyrinth.*

In this quote Galileo presents several of the basic tenets of modern western science. First, in order to understand or simulate nature, one must be conversant in its languages. Second, Nature's languages are mathematics and geometry is the specific language for describing, manipulating, and simulating shapes.

MAG 1.00E + 00   MAG 4.00E + 00   MAG 3.20E + 01

MAG 2.56E + 02   MAG 4.10E + 03   MAG 6.55E + 04

MAG 1.05E + 06   MAG 1.68E + 07   MAG 1.00E + 00

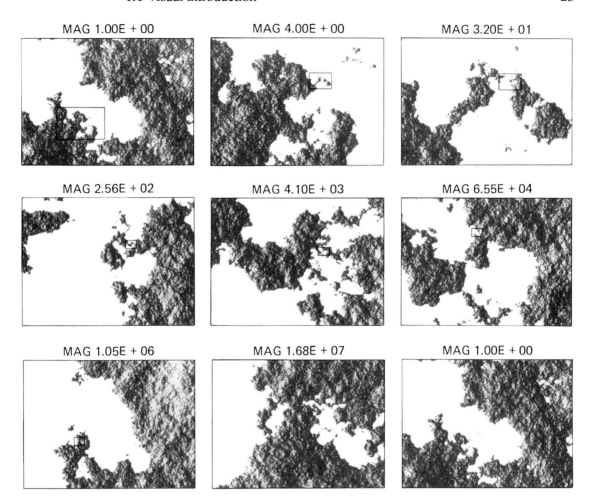

Fig. 1.1: Zoom sequence of the coastline of a statistically self-similar fractal landscape $D = 2.2$. Each succeeding picture shows a blowup of the framed portion of the previous image. As the surface is magnified, a small portion looks similar to (but not exactly the same as) a larger portion. The total magnification corresponds to 16 million.

Galileo was, however, wrong in terms of nature's preferred dialect. The inability of the triangles and circles of Euclidean geometry (the specific dialect mentioned) to concisely describe the natural world has not been widely appreciated until recently. Yet, in retrospect it is obvious that, "clouds are not spheres, mountains are not cones, coastlines are not circles, and bark is not smooth, nor does lightning travel in a straight line" (Mandelbrot, [68]). Fortunately, there is now a new dialect, a new branch of mathematics, that is appropriate for the irregular shapes of the real world, *fractal geometry* [68], as conceived and developed by Benoit Mandelbrot.

Fig. 1.2: Fractal Planetrise . A variation of Brownian motion on a sphere $D = 2.5$ rising above a fractally cratered random Gaussian fractal surface $D = 2.2$ (from Benoit Mandelbrot's *Fractal Geometry of Nature* [68]).

1.1.1 Mathematical monsters: The fractal heritage

The turn of the century coincided roughly with an upheaval in the world of mathematics. Minds conceived of strange monsters seemingly without counterpart in nature: Cantor sets, Weierstrass functions, Peano curves, curves without derivatives and lines that could fill space. Having once discovered these monsters (and congratulated themselves on a creativity superior to nature), mathematicians banished the pathological beasts, mostly sight unseen, to a mathematical zoo. They could imagine no use for, nor interest in, their creations by natural scientists. Nature, however, was not so easily outdone. As shown by Mandelbrot, many of these "monsters" do, in fact, have counterparts in the real world. Illuminated by computer graphics, they have finally been recognized as some of the basic structures in the language of nature's irregular shapes: the fractal geometry of nature.

Fractals (a word coined by Mandelbrot in 1975) have blossomed tremendously in the past few years and have helped reconnect pure mathematics research with both the natural sciences and computing. Within the last 5-10 years fractal geometry and its concepts have become central tools in most of the natural sciences: physics, chemistry, biology, geology, meteorology, and materials science. At the same time, fractals are of interest to graphic designers and filmmakers for their ability to create new and exciting shapes and artificial but realistic worlds. Fractal images appear complex, yet they arise from simple rules. Computer graphics has played an important role in the development and rapid acceptance of fractal geometry as a valid new discipline. The computer rendering of fractal shapes leaves no doubt of their relevance to nature. Conversely, fractal geometry now plays a central role in the realistic rendering and modelling of natural phenomena in computer graphics.

1.1.2 Fractals and self-similarity

What then are fractals? How are they different from the usual Euclidean shapes? Table 1.1 summarizes some of the major differences between fractals and traditional Euclidean shapes. First fractals are a decidedly modern invention. Although possessed of turn-of-the-century "pathological" ancestors, they have been recognized as useful to natural scientists only over the last 10 years. Second, whereas Euclidean shapes have one, or at most a few, characteristic sizes or length scales (the radius of a sphere, the side of a cube), fractals, like the coastline of Figure 1.1, possess no characteristic sizes. Fractal shapes are said to be self-similar and independent of scale or *scaling*. Third, Euclidean geometry provides concise accurate descriptions of man-made objects but is inappropriate for natural shapes. It yields cumbersome and inaccurate descriptions. It is likely that this limitation of our traditional language of shape is at least partly responsible for the striking qualitative difference between mass produced objects and natural shapes. Machine shops are essentially Euclidean factories: objects easily described are easily built. Fractals, on the other hand, provide an excellent description of many natural shapes and have already given computer imagery a natural flavor. Finally, whereas Euclidean shapes are usually described by a simple algebraic formula (e.g. $r^2 = x^2 + y^2$ defines a circle of radius r), fractals, in general, are the result of a construction procedure or algorithm that is often recursive (repeated over and over) and ideally suited to computers.

GEOMETRY

Mathematical language to describe, relate, and manipulate shapes

EUCLIDEAN	FRACTAL
• traditional (> 2000 yr)	• modern monsters ($\sim$ 10 yr)
• based on characteristic size or scale	• no specific size or scaling
• suits man made objects	• appropriate for natural shapes
• described by formula	• (recursive) algorithm

Table 1.1: A comparison of Euclidean and fractal geometries.

1.1.3   An early monster: The von Koch snowflake curve

These differences can can be illustrated with one of the early mathematical monsters: the von Koch snowflake curve (first proposed around 1904). Figure 1.3 illustrates an iterative or recursive procedure for constructing a fractal curve. A simple line segment is divided into thirds and the middle segment is replaced by two equal segments forming part of an equilateral triangle. At the next stage in the construction each of these 4 segments is replaced by 4 new segments with length $\frac{1}{3}$ of their parent according to the original pattern. This procedure, repeated over and over, yields the beautiful von Koch curve shown at the top right of Figure 1.3 .

It demonstrates that the iteration of a very simple rule can produce seemingly complex shapes with some highly unusual properties. Unlike Euclidean shapes, this curve has detail on all length scales. Indeed, the closer one looks, the more powerful the microscope one uses, the more detail one finds. More im-

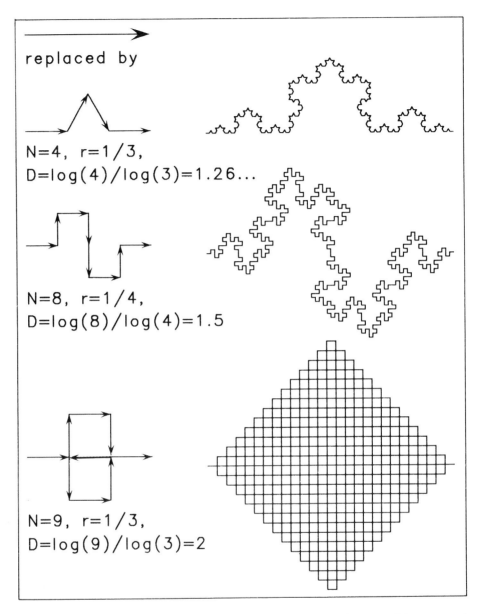

replaced by

N=4, r=1/3,
D=log(4)/log(3)=1.26...

N=8, r=1/4,
D=log(8)/log(4)=1.5

N=9, r=1/3,
D=log(9)/log(3)=2

Fig. 1.3: Recursive replacement procedure for generating the von Koch snowflake curve and variations with different fractal dimensions.

portant, the curve possesses an exact *self-similarity*. Each small portion, when magnified, can reproduce exactly a larger portion. The curve is said to be invariant under changes of scale much like the coastline of Figure 1.1. At each stage in its construction the length of the curve increases by a factor of $\frac{4}{3}$. Thus, the limiting curve crams an infinite length into a finite area of the plane without intersecting itself. As with the coastline, at successive iterations corresponding to successive magnifications, one finds new detail and increasing length. Finally, although the algorithm for generating the von Koch curve is concise, simple to describe, and easily computed, there is no algebraic formula that specifies the points of the curve.

1.1.4   Self-similarity and dimension

The property of self-similarity or *scaling*, as exemplified by the coastline, the von Koch curve, and the Mandelbrot set is one of the central concepts of fractal geometry. It is closely connected with our intuitive notion of *dimension* as illustrated in Figure 1.4 .

An object normally considered as one-dimensional, a line segment, for example, also possesses a similar scaling property. It can be divided into N identical parts each of which is scaled down by the ratio $r = \frac{1}{N}$ from the whole. Similarly, a two-dimensional object, such as a square area in the plane, can be divided into N self-similar parts each of which is scaled down by a factor $r = \frac{1}{\sqrt{N}}$. A three-dimensional object like a solid cube may be divided into N little cubes each of which is scaled down by a ratio $r = \frac{1}{\sqrt[3]{N}}$. With self-similarity the generalization to fractal dimension is straightforward. A D-dimensional self-similar object can be divided into N smaller copies of itself each of which is scaled down by a factor r where $r = \frac{1}{\sqrt[D]{N}}$ or

$$N = \frac{1}{r^D} \qquad (1.1)$$

Conversely, given a self-similar object of N parts scaled by a ratio r from the whole, its *fractal* or *similarity dimension* is given by

$$D = \frac{\log(N)}{\log(\frac{1}{r})} \qquad (1.2)$$

The fractal dimension, unlike the more familiar notion of Euclidean dimension, need not be an integer. Any segment of the von Koch curve is composed of 4 sub-segments each of which is scaled down by a factor of $\frac{1}{3}$ from its parent. Its fractal dimension is $D = \frac{\log(4)}{\log(3)}$ or about 1.26 .... This non-integer dimension, greater than one but less than two, reflects the unusual properties of the curve. It somehow fills more of space than a simple line ($D = 1$), but less than a Euclidean area of the plane ($D = 2$) . Mandelbrot [68] gives many variations of the von Koch construction and two others are presented in Figure 1.3 . In the middle a segment is replaced by 8 new segments each $\frac{1}{4}$ of the initial one to yield

$$D = \frac{\log(8)}{\log(4)} = 1.5 \ .$$

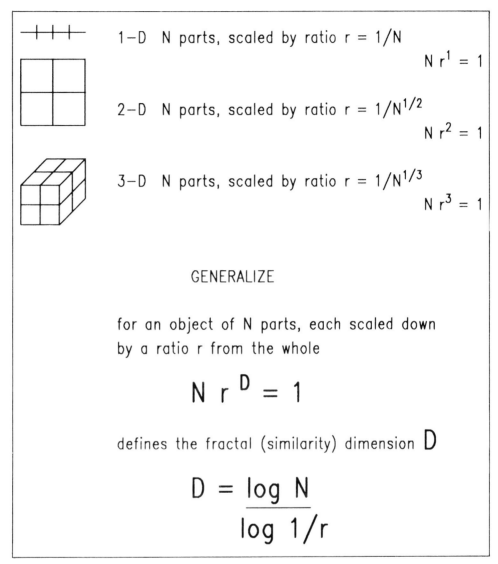

1-D   N parts, scaled by ratio r = 1/N

$$N r^1 = 1$$

2-D   N parts, scaled by ratio r = 1/N$^{1/2}$

$$N r^2 = 1$$

3-D   N parts, scaled by ratio r = 1/N$^{1/3}$

$$N r^3 = 1$$

GENERALIZE

for an object of N parts, each scaled down
by a ratio r from the whole

$$N r^D = 1$$

defines the fractal (similarity) dimension D

$$D = \frac{\log N}{\log 1/r}$$

Fig. 1.4: Interpretation of standard integer dimension figures in terms of exact self-similarity and extension to non-integer dimensioned fractals.

At the bottom each segment is replaced by 9 new segments each $\frac{1}{3}$ of the original for

$$D = \frac{\log(9)}{\log(3)} = 2 \ .$$

As D increases from 1 toward 2 the resulting "curves" progress from being "line-like" to "filling" much of the plane. Indeed, the limit $D \to 2$ gives a Peano or "space-filling" curve. The fractal dimension D, thus, provides a quantitative measure of wiggliness of the curves. Although these von Koch curves have fractal dimensions between 1 and 2, they all remain a "curve" with a topological dimension of one. The removal of a single point cuts the curve in two pieces.

1.1.5   Statistical self-similarity

The exactly self-similar von Koch curve may be considered a crude model for a coastline, but it differs from the coastline in one significant aspect. Upon magnification segments of the coastline look like, but never exactly like, segments at different scales. The concept of fractal dimension, however, can also be applied to such *statistically self-similar* objects. In a measurement of the length of a coastline, the more carefully one follows the smaller wiggles, the longer it becomes. A walk along a beach is longer than the drive along the corresponding coast highway. Moreover, each small section of a coastline looks like (but not exactly like) a larger portion. When using a ruler of size r to measure a coastline's length, the total length equals the ruler size r times the number of steps of size r, $N(r)$, taken in tracing the coast

$$Length = r \cdot N(r). \qquad (1.3)$$

As with the snowflake, $N(r)$ varies *on the average* as $\frac{1}{r^D}$ and

$$Length \propto r \cdot \frac{1}{r^D} = \frac{1}{r^{D-1}}. \qquad (1.4)$$

With $D > 1$, as the size of the ruler used to measure a coast decreases, its length increases. The variation of apparent coastline length with ruler size has been studied by Richardson as summarized in [68]. Real coastlines can, in fact, be characterized by fractal dimensions D of about 1.15 to 1.25, close to the $\frac{\log(4)}{\log(3)}$ of the von Koch curve.

The property that objects can look statistically similar while at the same time different in detail at different length scales, is the central feature of fractals in nature. The coastline of Figure 1.1 is random in the sense that (unlike the von Koch curve) a large scale view is insufficient to predict the exact details of a magnified view. Yet, the way in which the detail varies as one changes length scale is once again characterized by a fractal dimension. The irregular fractal surface is more than a simple surface ($D = 2$) and the actual surface shown in Figure 1.1 has $D = 2.2$. The fractal dimension of the coastline is one less than that of the surface itself. Here the coastline $D = 1.2$.

1.1.6   Mandelbrot landscapes

For the idealized fractal landscape of Figure 1.1 the statistical self-similarity extends from arbitrarily large to arbitrarily small scales. Actual landscapes, on the other hand, can be statistically self-similar only over a finite (but often quite large) range of distances. The largest variations may be limited by the size of the planet or the force of gravity (the materials may not be strong enough to support arbitrarily high mountains). The smallest scales may be limited by the smoothing of erosion, the basic grain size of the rock and sand or, at the very least, by the atomic nature of the particles. The mathematical ideal is an approximation to the real world (or vice versa to some mathematicians). Mandelbrot's fractal geometry, however, remains by far the best approximation and, to date, the most widely used successful mathematical model.

As discussed below, almost all algorithms for generating fractal landscapes effectively add random irregularities to the surface at smaller and smaller scales similar to the process of adding smaller and smaller line segments to the von Koch curve. Once again, the fractal dimension determines the relative amounts of detail or irregularities at different distance scales. Surfaces with a larger D seem rougher. This effect is illustrated in Plates 11 to 13 which show the "same" surface but with different fractal dimensions. Plate 11 has a relatively low fractal dimension for the surface, $D = 2.15$, that is appropriate for much of the earth. An increase in the fractal dimension to $D = 2.5$ is shown in Plate 12. A further increase to $D = 2.8$ in Plate 13 gives an unrealistically rough surface for an actual landscape. Fractal geometry specifies only the relative height variations of the landscape at different length scales. For the samples given here the color changes were based on height above water level and local slope.

The images in Plates 11 and 9 (which are the same except for the coloring) do not, however, resemble many of the valleys on the Earth's surface which are significantly eroded. A flat bottomed basin can, however, be approximated mathematically by simply taking the height variations (relative to water level) and scaling them by a power-law. This process is illustrated in Plate 10 where the original landscape of Plate 11 is "cubed." The effect of such a power greater than 1 is to flatten the lower elevations near the water emphasizing the peaks. Scaling with a power less than one has the opposite effect of flattening the peaks while increasing the steepness near the water. A cube-root processing of Plate 9 is shown in Plate 8. This forgery gives the impression of river erosion into an otherwise fairly smooth plain. Such non-linear processing does not change the fractal dimensions of the coastlines.

1.1.7   Fractally distributed craters

In order to give the impression of a lunar landscape, as shown in the foreground of Figure 1.2, it is necessary to add craters to the "virgin" fractal landscape of Plate 11. Each crater is circular with a height profile similar to the effect of dropping marbles in mud. The trick in achieving realism is to use the proper

Fig. 1.5: A fractal landscape (the same as Plate 11) with craters whose area distribution follows a fractal or power-law dependence with many more small craters than large ones.

distribution of crater sizes. For the moon, the actual distribution is fractal or power-law, with many more small craters than large ones. Specifically, the number of craters having an area, a, greater than some number A, $N(a > A)$ varies as $\frac{1}{A}$. The computer simulation is shown in Figures 1.5, 36 and on the back cover of this volume. This is the same surface that forms the foreground of Figure 1.2.

1.1.8   Fractal planet: Brownian motion on a sphere

Rather than add increasing detail at progressively smaller scales, the rising fractal planet in Figure 1.2 was generated by a different algorithm that is closer to an actual model for the evolution of the earth's surface. Its surface is a generalization of Brownian motion or a random walk on a sphere. A random walk is the sum of many independent steps, and the sphere's surface becomes the result of many independent surface displacements or faults. Each fault encircles the sphere in a random direction and divides it into two hemispheres which are displaced relative to each other in height. Inside the computer, the surface of the sphere is mapped onto a rectangular array similar to a flat projection map of the earth. Plate 1 shows the surface variations, as represented by regions of different color, both for the full and the flattened sphere after only 10 random faults. After 20 faults in Plate 2, the flattened sphere gives the impression of a cubist painting. After 100 faults in Plate 3 the land mass is scattered with an outline roughly resembling a rooster. With 500 faults in Plate 4, the sphere has developed a localized land mass in the northern hemisphere similar to the earth before continental drift. Moreover, the effect of an individual fault is becoming lost. After more than 10,000 faults and the addition of polar caps in Plate 5, the sphere's surface becomes a plausible planet with $D = 2.5$. Mapping the data of Plate 5 back onto a sphere gives the fractal planet in Figure 1.1 or the enlarged view of the rotated sphere shown in Plate 6. Remarkably, the same sphere is also a good forgery of a much more arid generic planet. Figure 7 shows the same data but with different coloring and without water. Any particular fractal model is often effective at simulating many different natural shapes. Once the fractal dimension is in an appropriate range, the mathematical models rarely fail to evoke natural images.

1.1.9   Fractal flakes and clouds

The previous fractal forgeries have been pretentious surfaces. The addition of irregularities on smaller and smaller scales raise their dimensions above a topological value of 2 to a fractal value $2 < D < 3$. It is, however, also possible to generate fractals with a topological dimension 3 whose scaling irregularities raise the fractal dimension to $3 < D < 4$. Rather than landscapes, which consist of random heights as a function of 2-dimensional surface position, such objects might represent a random temperature or water vapor density as a function of 3-dimensional position. The simplest means of viewing such constructions is with their *zerosets* (equivalent to the coastlines of the fractal landscapes) as shown in Figure 1.6. For a temperature distribution $T(x, y, z)$ with fractal di-

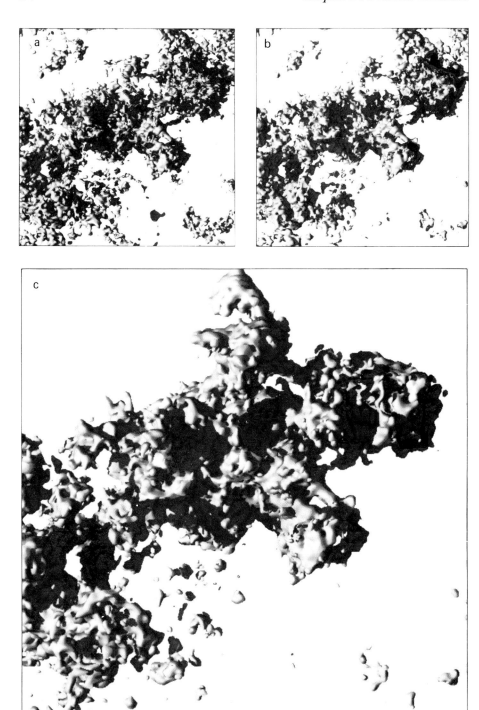

Fig. 1.6: Samples of fractal flakes with $D > 3$ as the zerosets of a fractal distribution $T(x, y, z) - T_0 = 0$ with **a.** $D - 1 = 2.7$, **b.** $D - 1 = 2.5$, **c.** $D - 1 = 2.2$.

mension $3 < D < 4$, the zerosets are all points for which $T(x, y, z) - T_0 = 0$ and have a fractal dimension of $D - 1$. They can be displayed by treating all points for which $T(x, y, z) > T_0$ as opaque while all others are transparent. The resulting fractal flakes with different D are shown in Figure 1.6 . In Figure 1.6(a) $D - 1$ is highest at 2.7 and the flake almost fills space. In Figure 1.6(b) $D - 1$ is reduced to 2.5 and the isolated shapes are beginning to condense. In Figure 1.6(c) $D - 1$ is reduced further to 2.2 and many natural shapes seem to appear. The overall outline is now reminiscent of a dog's head while just the upper portion could be the Loch Ness monster. Shapes with a fractal dimension D about 0.2 to 0.3 greater than the Euclidean dimension E seem particularly favored in nature. Coastlines typically have a fractal dimension around 1.2, landscapes around 2.2, and clouds around 3.3 . Even the large scale distribution of matter in the universe (clusters of clusters of galaxies) corresponds to D about 1.2 . Perhaps, then, it is not so surprising that our perceptual system, evolving in such a world, is particularly fond of $D - E$ about 0.2 .

Once again the generic nature of fractal constructions can be illustrated by allowing the local light scattering and opacity to vary continuously as $T(x, y, z)$. In this case, the flakes are transformed into realistic looking fractal clouds. A combination of such a fractal cloud above a cratered fractal surface is shown on the back cover. Here light scattered by the cloud produces shadows of varying intensity on the landscape. The correspondence to actual clouds is not surprising. Clouds and rain areas are the natural shapes where fractal behavior has been experimentally verified over the widest range of distance scales in [68].

1.2   Fractals in nature: A brief survey from aggregation to music

Man has always been confronted with a world filled with seemingly complex irregular shapes and random fluctuations. In the quest for understanding, natural science has progressed by concentrating primarily on the simplest of systems. In this process, it has moved away from the direct experience of nature to the electronically instrumented laboratory. After all, who could describe, let alone understand, the profile of a mountain or the shape of a cloud? With fractal geometry, the quest for scientific understanding and realistic computer graphic imagery can return to the everyday natural world. As the computer generated fractal forgeries [103] of Section 1.1 show, nature's complex shapes are only seemingly so.

1.2.1 Fractals at large scales

Diverse scientific disciplines have readily adopted the language of fractal geometry [15,36,61,85,95,97,108]. The images of Section 1.1 provide strong evidence of the applicability of fractals to nature on geological and planetary scales. Fractal geometry, however, also provides techniques for quantifying nature's irregular shapes, usually through a measurement of the fractal dimension [105]. A summary of these methods is given in the Section 1.6. At large scales, natural boundaries, geological topography, acid rain, cloud, rain and ecosystem boundaries, seismic faults, and the clustering of galaxies are all susceptible to fractal analysis.

1.2.2 Fractals at small scales: Condensing matter

Fractals are just as useful at smaller than human scales. The variation of surface roughness with D shown in Plates 11 to 13 suggests that D is also a good measurable quantifier for characterizing material surfaces on a microscopic scale. Materials scientists have widely adopted a fractal vocabulary for irregular shapes [108]. Fractals provide a general language for the taxonomy or classification of the various "animals" (previously pathological indescribable shapes) encountered in the natural sciences. Their ubiquity suggests two important questions: where do fractal shapes come from and how do the characteristics of fractal shapes effect the processes that occur on them ?

Many of the ways in which matter condenses on the microscopic scale seem to generate fractals. Figure 1.7 shows a transmission electron micrograph of a thin *gold film* [104] about 50 atomic layers thick on an amorphous substrate. Rather than a uniform coverage, the gold beads up due to surface tension (like water on a freshly waxed car hood). A computer analysis of the connected clusters (the largest is shown highlighted) shows irregular branching structures of finite size. As the amount of gold is increased, the clusters increase in size and eventually connect across the entire sample. As the largest cluster reaches the sample boundaries its shape is fractal over all length scales above the bead size. This is an example of a *percolation transition*, a geometric analog of a 2nd order phase transition. Here fractal geometry provides an alternative description to analytic scaling theory [104]. Fractal dimensions of the characteristic shapes correspond to the scaling exponents. Fractals also provide the language and the formalism for studying physical processes (such as diffusion and vibration) on such structures [81]. Diffusion on complex proteins with fractal structures has important biological implications.

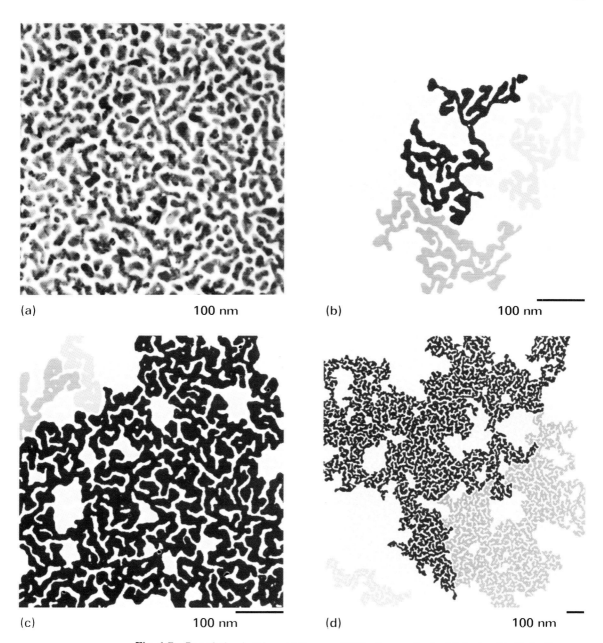

Fig. 1.7: Percolation in thin gold films. **a.** TEM micrograph of gold clusters (dark). **b.** Connectivity analysis of a. below threshold. **c.** Above threshold. **d.** At threshold.

Figures 1.8 and Plate 31 show samples of diffusion limited aggregation (see [36,95,97]). DLA is a simple model, easily implemented on a computer, that reproduces many natural shapes found in electrostatic discharges, electrochemical deposition, and fluid-fluid displacement. The resulting structures are well characterized as fractals and strongly reminiscent of plant-like growths. As with the Mandelbrot set, the resulting shapes are the result of repeated application of

simple non-linear rules. The DLA samples in Plate 31 begin with a single fixed
sticky particle at the origin of a sea of randomly moving particles. Each of these
mobile particles moves in its own random path with its own simple rules. At
each time step it moves one unit in a random direction unless it finds itself next
to the sticky origin. If so, it becomes immobile and serves as another sticky
site itself. Thus, beginning from one site, the cluster grows with the addition
of other particles that randomly reach its boundary. The open fjords never fill
since a particle trying to reach the bottom by a random path invariably hits one
of the sticky sides first. Figure 1.8 shows DLA samples with different initial
conditions (sticky line at the bottom) and changes of growth morphology with
sticking probability.

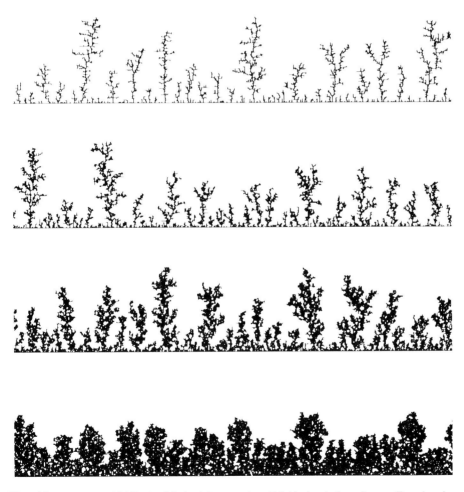

Fig. 1.8: Samples of Diffusion Limited Aggregation (DLA) simulations from a line showing
variations of growth from dendritic to moss-like.

1.2.3   Scaling randomness in time: $\frac{1}{f^\beta}$-noises

The preceding examples have demonstrated fractal randomness in space. Changes in time, however, have many of the same similarities at different scales as changes in space. To the physicist, unpredictable changes of any quantity V varying in time t are known as *noise*. Graphical samples of typical noises $V(t)$ are shown in Figure 1.9 . To the left of each sample is a representation of its *spectral densities*. The spectral density, $S_V(f)$, gives an estimate of the mean square fluctuations at frequency f and, consequently, of the variations over a time scale of order $\frac{1}{f}$. The traces made by each of these noises is a fractal curve and there is a direct relationship between the fractal dimension and the logarithmic slope of the spectral density which will be discussed below. Indeed, it is the understanding and simulation of such noises that forms the basis for all of the fractal forgeries presented here.

Figure 1.9(a) shows a *white noise*, the most random. It could be produced by a pseudo-random number generator and is completely uncorrelated from point to point. Its spectral density is a flat line, representing equal amounts at all frequencies (like a white light). Figure 1.9(c) shows a Brownian motion or a random walk, the most correlated of the three noise samples. It consists of many more slow (low frequency) than fast (high frequency) fluctuations and its spectral density is quite steep. It varies as $\frac{1}{f^2}$. Formally, the Brownian motion of Figure 1.9(c) is the integral of the white noise of Figure 1.9(a). In the middle, in Figure 1.9(b) is an intermediate type of noise known as $\frac{1}{f}$-noise because of the functional form of its spectral density. In general, the term $\frac{1}{f}$-noise is applied to any fluctuating quantity V(t) with $S_V(f)$ varying as $\frac{1}{f^\beta}$ over many decades with $0.5 < \beta < 1.5$. Both white and $\frac{1}{f^2}$-noise are well understood in terms of mathematics and physics. Although the origin of $\frac{1}{f}$-noise remains a mystery after more than 60 years of investigation, it represents the most common type of noise found in nature.

There are no simple mathematical models that produce $\frac{1}{f}$-noise other than the tautological assumption of a specific distribution of time constants. Little is also known about the physical origins of $\frac{1}{f}$, but it is found in many [102] physical systems: in almost all electronic components from simple carbon resistors to vacuum tubes and all semiconducting devices; in all time standards from the most accurate atomic clocks and quartz oscillators to the ancient hourglass; in ocean flows and the changes in yearly flood levels of the river Nile [64] as recorded by the ancient Egyptians; in the small voltages measurable across nerve membranes due to sodium and potassium flow; and even in the flow of automobiles on an expressway [78]. $\frac{1}{f}$-noise is also found in music.

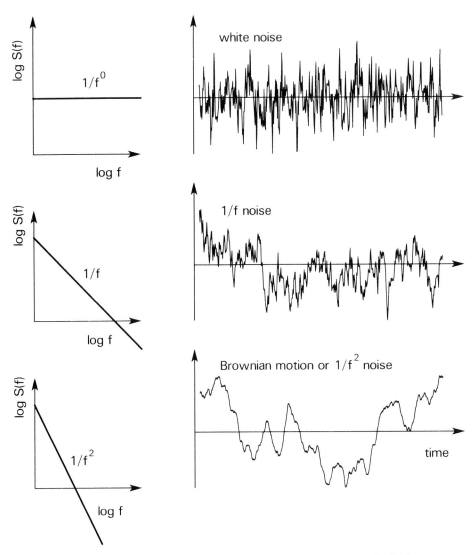

Fig. 1.9: Samples of typical "noises", $V(t)$, the random variations of a quantity in time.

a. White noise, the most random.

b. $\frac{1}{f}$-noise, an intermediate but very commonly found type of fluctuation in nature, its origin is, as yet, a mystery.

c. Brownian motion or a random walk.

To the left of each sample is a graphical representation of the spectral density, $S_V(f)$, a measurement characterizing the time correlations in the noise.

1.2.4 Fractal music

One of the most exciting discoveries [100,101] was that almost all musical melodies mimic $\frac{1}{f}$-noise. Music has the same blend of randomness and predictability that is found in $\frac{1}{f}$-noise. If one takes a music score and draws lines between successive notes of the melody one finds a graph remarkably similar in quality to Figure 1.9(b). Some of the actual measured melody spectral densities for different types of music is shown in Figure 1.10. There is little to distinguish these measurements on widely different types of music from each other or from the $\frac{1}{f}$-noise of Figure 1.9(b). Similar $\frac{1}{f}$-spectra are also observed for the loudness fluctuations in all types.

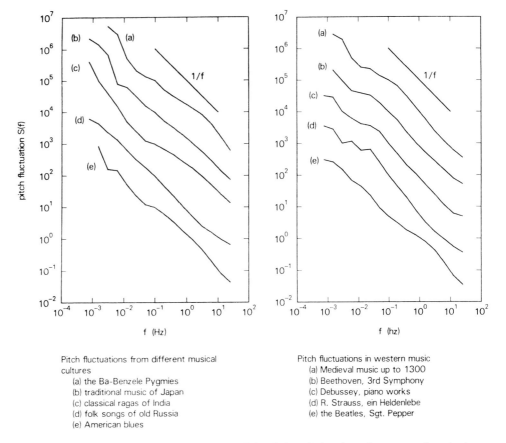

Pitch fluctuations from different musical cultures
   (a) the Ba-Benzele Pygmies
   (b) traditional music of Japan
   (c) classical ragas of India
   (d) folk songs of old Russia
   (e) American blues

Pitch fluctuations in western music
   (a) Medieval music up to 1300
   (b) Beethoven, 3rd Symphony
   (c) Debussey, piano works
   (d) R. Strauss, ein Heldenlebe
   (e) the Beatles, Sgt. Pepper

Fig. 1.10: Spectral density measurements of the pitch variations in various types of music showing their common correlations as 1/f-noise.

Measurement of spoken languages show a different result. Although the loudness fluctuations retain a $\frac{1}{f}$-spectrum, pitch fluctuations for spoken English show a Lorentzian behavior with white or independent variations for time scales longer than a spoken phoneme. Oriental languages, on the other hand, show an increase in spectral density of pitch fluctuations at low frequencies corresponding to increased correlation (but less that $\frac{1}{f}$). The $\frac{1}{f}$-spectral density seems to occur for measured quantities that are more closely related to "meaning" (musical melodies, spoken loudness).

This type of analysis is surprisingly insensitive to the different types of music. With the exception of very modern composers [100] like Stockhausen, Jolet, and Carter (where the melody fluctuations approach white noise at low frequencies), all types of music share this $\frac{1}{f}$-noise base. Such a view of melody fluctuations emphasizes the common element in music and suggests an answer to a question that has long troubled philosophers [45]. In the words of Plato, "For when there are no words (accompanying music), it is very difficult to recognize the meaning of the harmony and rhythm, or to see that any worthy object is imitated by them". Greek philosophers generally agreed on the imitative nature of the arts. It seemed obvious that painting, sculpture or drama imitated nature. But what does music imitate? The measurements suggest that music is imitating the characteristic way our world changes in time. Both music and $\frac{1}{f}$-noise are intermediate between randomness and predictability. Like fractal shapes there is something interesting on all (in this case, time) scales. Even the smallest phrase reflects the whole.

It is, of course, possible to use fractals, in this case as $\frac{1}{f^\beta}$-noises, for music as well as landscape forgery. Figure 1.11 shows samples of "music" generated from the three characteristic types of "noise" shown in Figure 1.9 . Although none of the samples in Figure 1.11 correspond to a sophisticated composition of a specific type of music, Figure 1.11(b) generated from $\frac{1}{f}$-noise is the closest to real music. Such samples sound recognizably musical, but from a foreign or unknown culture.

The above examples are only a small subset of the specialties in which fractals are finding application. Fractal geometry, like calculus before it, is rapidly moving from a specialized discipline in its own right to a common necessary tool and language in many fields. It remains important to emphasize the difference between the mathematical ideal and nature. Whereas mathematical fractals may provide scaling behavior over unlimited ranges, natural fractals always have limits. Nonetheless, fractal geometry remains the best available language for most natural shapes. As in most disciplines, an investigation of the boundaries (which are themselves often fractal) provides new insights.

1.3   Mathematical models: Fractional Brownian motion

One of the most useful mathematical models for the random fractals found in nature (such as mountainous terrain and clouds) has been the *fractional Brownian motion* (fBm) of Mandelbrot and Van Ness [68],[63]. It is an extension of the central concept of *Brownian motion* that has played an important role in both physics and mathematics. Almost all natural computer graphics fractal simulations [103] are based on an extension of fBm to higher dimensions such as the fractional Brownian landscape of Figure 1.1. Fractional Brownian motion

Fig. 1.11: Samples of stochastically composed fractal music based on the different types of noises shown in Figure 1.9 . **a.** "white" music is too random. **b.** "$\frac{1}{f}$"-music is the closest to actual music (and most pleasing). **c.** "Brown" or $\frac{1}{f^2}$. Music is too correlated.

is also a good starting point for understanding anomalous diffusion and random walks on fractals. This section provides a summary of the usage of fBm. Its mathematical details are left to the Section 1.6 .

As can be seen from the sample *traces* of fBm in Figure 1.12, a fractional Brownian motion, $V_H(t)$, is a single valued function of one variable, t (usually time). In appearance, it is reminiscent of a mountainous horizon or the fluctuations of an economic variable. Formally, it is the increments of fBm (the differences between successive times) that give the noises of Figure 1.9. The scaling behavior of the different traces in Figure 1.12 is characterized by a parameter H in the range $0 < H < 1$. When H is close to 0 the traces are roughest while those with H close to 1 are relatively smooth. H relates the typical change in V, $\Delta V = V(t_2) - V(t_1)$, to the time difference $\Delta t = t_2 - t_1$ by the simple scaling law

$$\Delta V \propto \Delta t^H. \tag{1.5}$$

In the usual Brownian motion or random walk, the sum of independent incre-

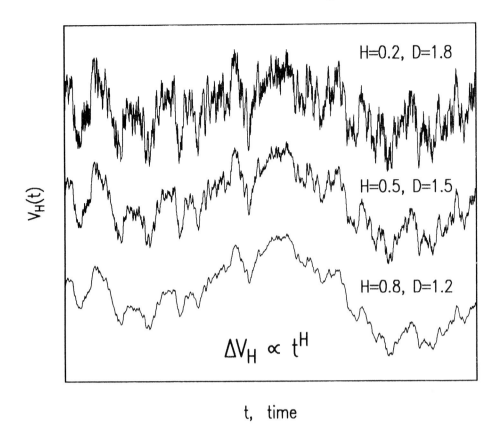

Fig. 1.12: Samples of fractional Brownian motion traces $V_H(t)$ vs t for different values of H and D.

ments or steps leads to a variation that scales as the square root of the number of steps. Thus, $H = \frac{1}{2}$ corresponds to a trace of Brownian motion.

1.3.1 Self-affinity

The scaling property of fBm represented by Eqn. (1.5) is different from the statistical self-similarity of the coastline of Figure 1.1 and the exact self-similarity of the Koch constructions of Figure 1.3. Whereas the self-similar shapes repeat (statistically or exactly) under a magnification, the fBm traces of Figure 1.12 repeat statistically only when the t and V directions are magnified by *different* amounts. If t is magnified by a factor r (t becomes rt), then V must be magnified by a factor r^H (V becomes $r^H V$). For a random walk ($H = \frac{1}{2}$) one must take four times as many steps to go twice as far. This non-uniform scaling, where shapes are (statistically) invariant under transformations that scale different coordinates by different amounts, is known as *self-affinity*. Self-affinity also plays an important role in the properties of iterated functions as discussed in Chapter 5 .

As shown in Figure 1.4, the concept of fractal dimension is strongly connected with self-similar scaling. Although the extension to self-affine shapes can be ambiguous (see Section 1.6), a useful and consistent definition is based on the concept of *zerosets* and the calculus of fractal dimensions [68],[71].

1.3.2 Zerosets

Fractals, like traditional Euclidean shapes, typically reduce their dimension by one under intersection with a plane. Thus, the intersection of a solid 3-d sphere with a plane is a 2-d circular area. The intersection of this area with another plane is a 1-d line segment and the intersection of this segment with yet another plane is a 0-d point. Similarly, the intersection of a fractal curve in the plane (with fractal dimension $1 < D < 2$) with a straight line is a set of points of dimension $D - 1$. By choosing the direction of the intersecting line to eliminate one of the coordinates, it is possible to reduce a self-affine curve to a self-similar set of points. The *zeroset* of fBm is the intersection of the trace of $V_H(t)$ with the t axis: the set of all points such that $V_H(t) = 0$. The zeroset is a disconnected set of points with topological dimension zero and a fractal dimension $D_0 = 1 - H$ that is less than 1 but greater than 0. Although the trace of $V_H(t)$ is self-affine, its zeroset is self-similar and different estimates of D_0 yield the same answer. The fractal dimension, $D = D_0 + 1$, of a self-affine fBm is, thus, simply related to the scaling parameter H as

$$D = 2 - H. \qquad (1.6)$$

1.3.3 Self-affinity in higher dimensions : Mandelbrot landscapes and clouds

The traces of fBm, particularly Figure 1.12(c) with $H = 0.8$, bear a striking resemblance to a mountainous horizon. The modelling of the irregular Earth's surface as a generalization of traces of fBm was first proposed by Mandelbrot. The single variable t can be replaced by coordinates x and y in the plane to give $V_H(x, y)$ as the surface altitude at position x, y as shown in Figure 1.1. In this case, the altitude variations of a hiker following any straight line path at constant speed in the xy-plane is a fractional Brownian motion. If the hiker travels a distance Δr in the xy-plane ($\Delta r^2 = \Delta x^2 + \Delta y^2$) the typical altitude variation ΔV is given by

$$\Delta V \propto \Delta r^H \qquad (1.7)$$

in analogy with Eqn. (1.5). Once again, the fractal dimension D must be greater than the topological dimension 2 of the surface. Here,

$$D = 3 - H \quad \text{for a fractal landscape} \quad V_H(x, y). \qquad (1.8)$$

The intersection of a vertical plane with the surface $V_H(x, y)$ is a self-affine fBm trace characterized by H and has a fractal dimension one less than the surface itself.

Similarly, the zeroset of $V_H(x, y)$, its intersection with a horizontal plane, has a fractal dimension $D_0 = 2 - H$. This intersection, which produces a family of (possibly disconnected) curves, forms the coastlines of the $V_H(x, y)$ landscape. Since the two coordinates x and y are, however, equivalent, the coastlines of the self-affine $V_H(x, y)$ are self-similar. Figure 1.1 demonstrates how the self-similar coastline remains statistically invariant under changes of magnification. A side view of the same landscape would not, however, show the same invariance because of its self-affine scaling. Viewed from nearby a fractal landscape such as the earth's surface appears relatively steep. On the other hand, when the same section is viewed from further away (for example, a satellite view of the earth) the surface is almost flat. This is just a consequence of the scaling property of Eqn. (1.7). If the close-up corresponds to ΔV of order Δr then increasing the viewing distance by a factor $F > 1$ corresponds to increasing the horizontal scale by F and the vertical scale by only F^H. Since $H < 1$ the vertical scale increases by less than the horizontal and the surface appears to flatten. The relative height variations seen in a fractal landscape (which may be varied arbitrarily in a simulation) are an important perceptual clue in estimating the height of the mountains relative to human scale.

This generalization of fBm can continue to still higher dimensions to produce, for example, a self-affine fractal temperature or density distribution $V_H(x, y, z)$. Here, the variations of an observer moving at constant speed along any straight line path in xyz-space generate a fBm with scaling given by Eqn. (1.7) where now $\Delta r^2 = \Delta x^2 + \Delta y^2 + \Delta z^2$. In analogy with Eqs. (1.6) and (1.8), the fractal dimension

$$D = 4 - H \quad \text{for a fractal cloud} \quad V_H(x, y, z). \qquad (1.9)$$

The zeroset

$$V_H(x, y, z) = \text{constant}$$

now gives a self-similar fractal with $D_0 = 3 - H$ as the fractal flakes of Figure 1.6.

To summarize, a statistically self-affine fractional Brownian function, V_H, provides a good model for many natural scaling processes and shapes. As a function of one variable ($E = 1$) fBm is a good mathematical model for noises, random processes, and music. For $E = 2$ fBm provides fractal landscapes and random surfaces. For $E = 3$ fBm generates flakes and clouds. In all of these cases, the scaling property may be characterized equivalently by either H or the fractal dimension D. Yet another characterization is provided by the *spectral exponent* β.

1.3.4 Spectral densities for fBm and the spectral exponent β

As discussed earlier, random functions in time $V(t)$ are often characterized [90],[91] by their *spectral densities* $S_V(f)$. If $V(t)$ is the input to a narrow bandpass filter at frequency f and bandwidth Δf , then $S_V(f)$ is the mean square of the output $V(f)$ divided by Δf, $S_V(f) = \frac{|V(f)|^2}{\Delta f}$. $S_V(f)$ gives information about the time correlations of $V(t)$. When $S_V(f)$ increases steeply at low f, $V(t)$ varies more slowly. As shown in Figure 1.9, the spectral exponent β changes with the appearance of the noise trace similar to the variation with H of the traces of fBm from Figure 1.12. The Sections 1.6.8 and 2.4 provide a useful connection between the 3 equivalent characterizations, D, H, and β, of a fBm function of E variables

$$D = E + 1 - H = E + \frac{3 - \beta}{2}. \qquad (1.10)$$

For H in the range $0 < H < 1$, D spans the range $E < D < E + 1$, and $1 < \beta < 3$. The value $H \simeq 0.8$ is empirically a good choice for many natural phenomena.

1.4   Algorithms: Approximating fBm on a finite grid

Section 1.1 established the visual connection between many, seemingly complex, shapes in the natural world and statistically self-similar and self-affine fractals. Section 1.2 indicated some of the ways in which fractals are becoming an integral part of natural science. Section 1.3 reviewed some of the mathematical background and established the relation between fractal dimension D, the H parameter of fBm, and the spectral density exponent β. This section discusses various methods for producing a finite *sample* of fBm as a noise ($E = 1$), a landscape ($E = 2$), or a cloud ($E = 3$). As with nature itself, this sample will typically be limited by both a smallest size or resolution L_{min} and a largest scale L_{max}. Consideration will be given to sample statistics (mathematically, how well does it approximate fBm?), visual features (does it look natural?) and computation characteristics (how does computation time vary with sample size?). In general, the closer a given algorithm approximates fBm, the more "realistic" the resulting image looks. Although no specific consideration will be given to image rendering, it is important to note that in most cases rendering of a fractal sample requires far more computation than sample generation. This is particularly true for rendering packages based on Euclidean shapes which have an extremely difficult time with the "infinite" number of surface patches on a fractal surface. The Chapter 2 in this volume also provides a practical and detailed account of some of these algorithms along with pseudo-code samples.

1.4.1 Brownian motion as independent cuts

The earliest simulations of fBm were based on the idea of independent perturbations. In the usual definition, Brownian motion, $V_B(t) = V_H(t)$ for $H = \frac{1}{2}$, is the sum of independent increments or pulses. Formally, $V_B(t)$ is the integral of a Gaussian white noise W(t),

$$V_B(t) = \int_{-\infty}^{t} W(s)\,ds. \qquad (1.11)$$

Since a single pulse at time t_i causes a step-function jump of magnitude A_i (a Gaussian random variable) in $V(t)$, Brownian motion may equivalently be considered as the cumulative displacement of a series of independent jumps. The response to such a pulse, $A_i P(t - t_i)$ has the form of a step function with $P(t) = 1$ for $t > 0$ and $P(t) = 0$ for $t \leq 0$. Thus, $V_B(t)$ can also be written as the sum of independent cuts at random (Poisson distributed) times t_i

$$V_B(t) = \sum_{i=-\infty}^{\infty} A_i P(t - t_i). \qquad (1.12)$$

This latter formulation of Brownian motion is useful since it can be generalized to circles and spheres to produce the fractal planet of Plate 6. In the generalization to a sphere, V_B becomes a function of $\vec{r}$, the position on the unit sphere. Local variations of V_B mimic the usual Brownian motion but large scales reflect the periodicity of the sphere. This corresponds to the addition of random hemispheres whose pole positions $\vec{r_i}$ are uniformly distributed on the sphere surface,

$$V_B(\vec{r}) = \sum_{i=-\infty}^{\infty} A_i P(\vec{r} - \vec{r_i}). \qquad (1.13)$$

Here, $P(\vec{r} - \vec{r_i}) = 1$ for $\vec{r} \cdot \vec{r_i} > 0$ and zero otherwise. The evolution of a Brownian sphere under this summation process is shown in Plates 1 to 6.

A flat Brownian relief $V_B(x, y)$ can similarly be constructed from the addition of randomly placed and randomly oriented faults in the plane. The profile of such a fault corresponds to the step function of random amplitude A_i. Such a surface will have a fractal dimension $D = 2.5$.

The independent addition of such step function faults is extremely expensive computationally. Each fault requires additions to roughly half of the surface elements. Moreover, the resulting fractal always corresponds to a fBm with $H = \frac{1}{2}$. Nevertheless, the procedure is straightforward and represents, historically, the first method used for producing fractal landscapes [68].

By changing the profile of the cut from the step function it is possible to use the method of independent cuts to generate fractal surfaces and planets with different H and D. Thus,

$$P(t) = t^{H - \frac{1}{2}} \quad \text{for} \quad t > 0$$

and

$$P(t) = - |t|^{H - \frac{1}{2}} \quad \text{for} \quad t < 0$$

can be used to generate a fBm with H different from $\frac{1}{2}$.

1.4.2 Fast Fourier Transform filtering

Another straightforward algorithm for fBm is to directly construct a random function with the desired spectral density $\frac{1}{f^\beta}$. β is related to the desired D or H by Eqn. (1.10) . For all practical purposes, the output of a pseudo-random number generator produces a "white noise" $W(t)$ with constant spectral density $S_W(f)$. Filtering $W(t)$ with a transfer function $T(f)$ produces an output, $V(t)$, whose spectral density,

$$S_V(f) \propto |T(f)|^2 S_W(f) \propto |T(f)|^2 .$$

Thus, to generate a $\frac{1}{f^\beta}$-noise from a $W(t)$ requires

$$T(f) \propto \frac{1}{f^{\frac{\beta}{2}}}.$$

For computer simulation, a continuous function of time, $V(t)$, is approximated by a finite sequence of N values, V_n, defined at discrete times $t_n = n\Delta t$, where n runs from 0 to $N - 1$ and Δt is the time between successive values. The discrete Fourier transform (DFT) defines V_n in terms of the complex Fourier coefficients, v_m , of the series:

$$V_n = \sum_{m=0}^{\frac{N}{2} - 1} v_m e^{2\pi i f_m t_n}, \tag{1.14}$$

where the frequencies

$$f_m = \frac{m}{N\Delta t} \quad \text{for} \quad m = 0 \quad \text{to} \quad \frac{N}{2} - 1.$$

For a fBm sequence with $S_V(f)$ varying as $\frac{1}{f^\beta}$ the coefficients must satisfy[1]

$$<| \, v_m \, |^2> \propto \frac{1}{f^\beta} \propto \frac{1}{m^\beta} \qquad\qquad (1.15)$$

The relation between β and D and H is given by Eqn. (1.10) with $E = 1$. The v_m may be obtained by multiplying the Fourier coefficients of a white noise sequence by $\frac{1}{f^{\frac{\beta}{2}}}$ or by directly choosing complex random variables with mean square amplitude given by Eqn. (1.15) and random phases. One possible variation sets

$$| \, v_m \, | = \frac{1}{m^{\frac{\beta}{2}}}$$

and only randomizes the phases. With an FFT (Fast Fourier Transform) algorithm the evaluation of Eqn. (1.14) requires of order $N \log N$ operations [17] to produce a series of N points. As such it offers a significant improvement over the random cuts described above.

Since the FFT includes all possible frequency components over the range of distance scales from L_{min} to $L_{max} = N \cdot L_{min}$ it represents an accurate approximation to fBm. It has, however, several drawbacks. The entire sample must be computed at once and it is difficult to vary the degree of detail across the sample or to extend the sample across its original boundaries. In addition, the result is periodic in time, $V_n = V_{n+N}$. Such boundary constraints become more obvious as $\beta \to 3$, $H \to 1$, and $D \to 1$. This artifact may be reduced by generating a longer sequence and keeping only a portion (typically $\frac{1}{4}$ to $\frac{1}{2}$). Nevertheless, the FFT transform has been used to produce some of the most striking still images of random fractals (as shown in the color plates). In most cases the rendering takes significantly longer than the generation of the fractal data base itself.

The procedure is easily extended to spatial fBm functions of E coordinates, $V_H(\vec{x})$, by scaling the corresponding spatial frequencies $\vec{k}$. Once again the fBm function is defined as the Fourier series of its spatial frequency components

$$V_H(\vec{x}) = \sum v_k e^{2\pi i \vec{k} \cdot \vec{x}} \qquad\qquad (1.16)$$

where the scaling of the spatial frequencies determines D, H, and β. In the Section 1.6 it is shown that a fractional Brownian surface ($E = 2$) corresponds

---

[1] The brackets $<>$ denote averages over many samples.

to

$$\beta = 1 + 2H = 7 - 2D$$

and

$$v_k \propto \frac{1}{|k|^{\beta+1}} \propto \frac{1}{(k_x^2 + k_y^2)^{4-D}} \qquad (1.17)$$

with $2 < D < 3$. Although neither the random cuts nor the FFT filtering represent a *recursive* method of generating random fractals, the process of adding more and more frequencies is similar to the construction of the von Koch snowflake. Figure 1.13 shows the increasing complexity in the evolution of an FFT generated fBm surface with $H = 0.8$ and $D = 2.2$ as higher frequencies are included.

Similarly, for a fractional Brownian volume ($E = 3$)

$$v_k \propto \frac{1}{|k|^{\beta+2}} \propto \frac{1}{(k_x^2 + k_y^2 + k_z^2)^{\frac{11-2D}{2}}} \qquad (1.18)$$

with $3 < D < 4$. This method was used to produce the fractal flakes and clouds of Figure 1.6 and on the back cover.

1.4.3 Random midpoint displacement

Random midpoint displacement is a recursive generating technique that was applied to normal Brownian motion as early as the 1920's by N.Wiener. It is a natural extension of the von Koch construction and figures in many of the fractal samples described by Mandelbrot [68]. Its use in computer graphics has been widely popularized by Fournier, Fussell, and Carpenter [19,40]. For $H \neq \frac{1}{2}$ or $E > 1$, it sacrifices mathematical purity for execution speed in its approximation to fBm.

Consider the approximation to a simple fBm, $V_H(t)$, where the mean square increment for points separated by a time $\Delta t = 1$ is σ^2. Then, from Eqn. (1.5), for points separated by a time t,

$$<|V_H(t) - V_H(0)|^2> = t^{2H} \cdot \sigma^2. \qquad (1.19)$$

If, for convenience, $V_H(0) = 0$, then the points at $t = \pm 1$ are chosen as samples of a Gaussian random variable with variance σ^2 to satisfy Eqn. (1.5). Given these initial conditions, one defines the midpoints at

$$V_H(\pm\frac{1}{2}) = 0.5 [V_H(0) + V_H(\pm 1)] + \Delta_1,$$

The first term is the desired total variance from Eqn. (1.5) while the second term represents the fluctuations already in

$$\Delta V_H(1) = V_H(\pm 1) - V_H(0)$$

due to the previous stage. As $H \to 1, \Delta_1^2 \to 0, D \to 1$, no new fluctuations are added at smaller stages, and $V_H(t)$ remains a collection of smooth line segments connecting the starting points. At the second stage,

$$V_H(\pm \frac{1}{4}) = 0.5 \left[V_H(0) + V_H(\pm \frac{1}{2}) \right] + \Delta_2$$

where Δ_2 has variance

$$\Delta_2^2 = \frac{\sigma^2}{4^{2H}} - \frac{1}{4} \text{var} \left[\Delta V_H(\frac{1}{2}) \right] = \frac{\sigma^2}{4^{2H}} \left[1 - 2^{2H-2} \right].$$

At the n th stage, the length scale has decreased to $\frac{1}{2^n}$ and a random Gaussian variable Δ_n is added to the midpoints of the stage $n-1$ with variance

$$\Delta_n^2 = \frac{\sigma^2}{(2^n)^{2H}} \left[1 - 2^{2H-2} \right] \qquad (1.20)$$

As expected for a fBm, at a length scale $r = \frac{1}{2^n}$ one adds randomness with mean square variations varying as r^{2H}.

Although this process does produce a fractal, the result is, unfortunately, not stationary [68,69] for all H. Once a given point at t_i has been determined, its value remains unchanged in all later stages. All additional stages change $t < t_i$ independent from $t > t_i$ and the correlations required of fBm with $H \neq \frac{1}{2}$ are not present. By construction the increments

$$<| V_H(\pm 1) - V_H(0) |^2 >= \sigma^2.$$

For a stationary process, the same should be true of all increments with $\Delta t = 1$. However, the absence of correlation across an earlier stage requires that

$$<| V_H(\frac{1}{2}) - V_H(-\frac{1}{2}) |^2 >= 2 <| V_H(\frac{1}{2}) - V_H(0) |^2 >= 2 \frac{\sigma^2}{2^{2H}}.$$

This gives the desired result σ^2 only for the $H = \frac{1}{2}$ of normal Brownian motion.

As a consequence, points generated at different stages have different statistical properties in their neighborhoods. This often leaves a visible trace that does not disappear as more stages are added. The effect is more pronounced as $H \to 1$. These visible artifacts, which are a consequence of the lack of stationarity of the mathematical approximation, are particularly visible on fractal surfaces. Figure 1.14 shows a zenith view of such a midpoint displacement surface with $H = 0.8$. In the generation of a midpoint displacement surface on a square grid each step proceeds in two stages. First the midpoints of each of the squares is determined from its 4 corners and shifted by a random element Δ. This determines a new square lattice at 45 degrees to the original and with lattice size $\frac{1}{\sqrt{2}}$. In the second stage, the midpoints of the new lattice receive a random contribution smaller by $\frac{1}{\sqrt{2^H}}$ from the first stage. This produces the new square lattice with a scale $\frac{1}{2}$ the original. The traces of early stages are readily visible in Figure 1.14 . These artifacts, which occur at all stages, can not be eliminated by local smoothing.

In spite of its mathematical failings, the speed of midpoint displacement, and its ability to add "detail" to an existing shape make it a useful fractal algorithm for some applications. To generate N points requires only order N operations. To extend the sequence by just one point at the smallest scale beyond its original endpoints, however, requires an additional $\log_2 N$ operations.

1.4.4   Successive random additions

In many respects the non-stationary artifacts of midpoint displacement are similar to the staircase effect of aliased raster display lines. With midpoint displacement, once determined, the value at a point remains fixed. At each stage only half of the points are determined more accurately. If one imagines the process of magnifying an actual object, as the spatial resolution increases *all points are determined more accurately*. In terms of the Nyquist sampling theorem, to approximate N real points requires $\frac{N}{2}$ complex frequencies or $\frac{N}{2}$ sine and cosine components. When the resolution is doubled to $2N$ points, the additional high frequency components alter all of the original values. Midpoint displacement only adds the additional sine (or cosine) components, not both. Conversely, in reducing the resolution of an image an anti-aliasing procedure will average over neighboring pixels. Simply keeping the same value as the center (the equivalent of midpoint displacement) produces the objectionable staircase edge.

I call the process of adding randomness *to all points* at each stage of a recursive subdivision process *successive random additions* . This enhancement reduces many of the visible artifacts of midpoint displacement and the generation still requires only order N operations to generate N points. The computation of the midpoints is the same as midpoint displacement. The only difference is in the number of random elements. For a sequence of N elements, $\frac{N}{2}$ points in the final stage had only one random addition. $\frac{N}{4}$ points in the previous stage had 2 random additions. $\frac{N}{8}$ had 3 and so on. The series converges to the $2N$ random additions for the N elements.

The zoom sequence of Figure 1.1 was produced with successive random additions. The artifacts of the square lattice are not as visible as with the midpoint displacement surface of Figure 1.14 . For comparison Figure 1.15 also shows the construction of a fBm surfaces with $H = 0.8$ by successive random additions.

With successive random additions, at each stage all points are treated equivalently. This has the additional advantage that the resolution at the next stage can change by any factor $r < 1$. For midpoint displacement r must be $\frac{1}{2}$. Thus, given a sample of N_n points at stage n with resolution l, stage $n + 1$ with resolution rl is determined by first interpolating the $N_{n+1} = \frac{N_n}{r}$ new points from the old values. In practice, this can be accomplished using either linear or spline interpolation. A random element Δ_n is added to all of the new points. At stage n with scaling ratio $r < 1$, the Δ will have a variance

$$\Delta_n^2 \propto \left(r^n \right)^{2H} . \tag{1.21}$$

When $\frac{1}{r}$ is an integer, the generation of a sequence of N points requires order $C(r)N$ operations. The coefficient $C(r)$ varies as $\sum n(1 - r)^n$ over the number of stages.

The fractal dimension D for the generated objects is determined only by H, r can be varied independently. Variations in r change the *lacunarity* of the fractal. Figure 1.15 shows two samples of successive random addition surfaces (both with $H = 0.8, D = 2.2$) generated from differing r. The zoom sequence in Figure 1.1 and Figure 1.15(a) were generated with $\frac{1}{r} = 2$. As $\frac{1}{r}$ increases to 8 in Figure 1.15(b) the few characteristic resolutions at which randomness has been added become visible and the lacunarity increases. If $\frac{1}{r}$ approaches 1 the surfaces approach the characteristics of the FFT samples. Empirically, however, a value of $\frac{1}{r}$ much smaller than 2 produces little observable change. The free choice of a value for r is an important addition for filmmaking. The addition of irregularities to a surface can be accomplished continuously as the resolution is slowly increased from frame to frame.

1.4.5 Weierstrass-Mandelbrot random fractal function

Each stage in a midpoint displacement process increases the highest spatial frequency by a factor $\frac{1}{r} = 2$. Each stage of successive random additions increases this frequency by $\frac{1}{r} > 1$. Thus, both are related to Mandelbrot's generalization [11,68] of the Weierstrass non-differentiable function, $V_{MW}(t)$. Whereas the

Fig. 1.14: Zenith and tilted views of a midpoint displacement surface on a square lattice for $H = 0.8$ and $D = 2.2$. The illumination was along one lattice direction. The non-stationary character is visible as the prominent shadows. This may be compared with Plates 11 to 13 and Figure 1.15 which were also generated on a square lattice.

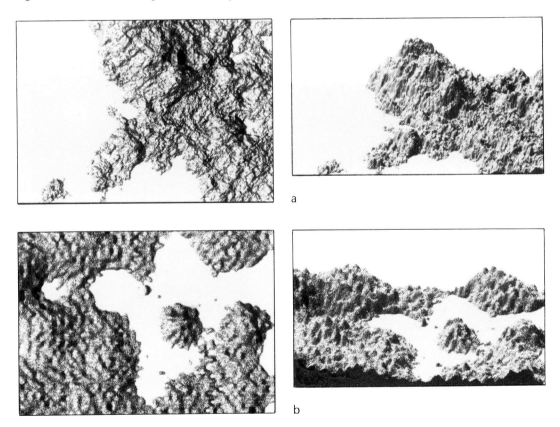

a

b

Fig. 1.15: Zenith and tilted views of successive random addition surfaces on a square lattice for $H = 0.8$ and $D = 2.2$. The illumination was along one of the axes. By varying the expansion factor $\frac{1}{r}$ before new randomness is added, the surface *lacunarity* (or texture) can be varied without changing D. **a.** $\frac{1}{r} = 2$. **b.** $\frac{1}{r} = 8$. The lacunarity increases as $\frac{1}{r}$ increases and for large $\frac{1}{r}$ only a few fluctuation scales are visible. As $r \rightarrow 1$ the lacunarity approaches that of the FFT surfaces. A comparison of these surfaces with the midpoint displacement surface of Figure 1.14 shows how the addition of random components to all points at each magnification eliminates most of the visible artifacts.

Fourier series of Eqn. (1.14) involves a *linear* progression of frequencies, the Weierstrass function involves a geometric progression.

$$V_{MW}(t) = \sum_{n=-\infty}^{\infty} A_n r^{nH} \sin\left(2\pi r^{-n} t + \phi_n\right), \qquad (1.22)$$

where A_n is a Gaussian random variable with the same variance for all n and ϕ_n is a random phase uniformly distributed on $[\,0\,,2\,\pi\,]$. The original Weierstrass function did not include the terms for $n < 0$ which add small spatial frequencies and large scale fluctuations. As with successive random additions, the addition of a new term to the V_{MW} sum decreases the spatial resolution by r and adds new fluctuations with variance $A^2 r^{2H}$. In terms of spectral density, although V_{MW} contains only discrete frequencies $f_n = \frac{1}{r^n}$, each component has a mean square amplitude proportional to r^{2Hn} or $1/f_n^{2H}$ in a bandwidth $\Delta f \propto f$. Thus, the spectral density

$$S_{V_{MW}}(f) \propto \frac{amplitude^2}{\Delta f} \propto \frac{1}{f^{2H+1}} \qquad (1.23)$$

in agreement with Eqn. (1.39).

Although the V_{MW} sum involves an infinite number of components, all practical applications introduce both low and high frequency cutoffs. For a range of distance scales from L_{min} to L_{max} only log $\frac{L_{max}}{L_{min}}$ components are needed. Frequencies much higher than $\frac{1}{L_{min}}$ average out to zero contribution over scales of size L_{min} while frequencies much lower than $\frac{1}{L_{max}}$ contribute only an overall offset and slope. The V_{MW} random fractal function of Eqn. (1.22) allows the direct computation of $V_{MW}(t)$ *at any* t from a stored set of order log $\frac{L_{max}}{L_{min}}$ coefficients A_n and ϕ_n.

With the use of table lookup for $\sin(x)$, the V_{MW} fractal function can be made extremely fast. The fractal is represented by a only few stored coefficients and there is no penalty for calculating the function outside its original boundary. The resolution can easily be changed by changing the number of components used. A variable r allows changing lacunarity independent of D. Other periodic functions can be used in place of $\sin(x)$. For example, both midpoint displacement and successive random additions, correspond to a triangle function. In addition, midpoint displacement sets all ϕ_n = constant. Extensions to $E > 1$ are possible with periodic functions in all E coordinates.

1.5 Laputa: A concluding tale

Jonathan Swift in his 1726 classic describes how Capt. Gulliver, captured by pirates, set adrift, and marooned on a small island, is eventually rescued by a

chain lowered from the large floating island of Laputa. His sense of relief, however, turns quickly to dismay. His rescuers, the Laputans, have but two interests:
music and mathematics. So exclusive and intense is their speculation that they
have effectively withdrawn from the normal everyday world and must employ
servants, known as *flappers*, to strike them upon the ear or mouth (rather like a
primitive form of I/O interrupt) whenever normal social intercourse is required.
Their obsession, moreover, extends to all spheres. For his first meal Gulliver
is served "a shoulder of mutton, cut into an equilateral triangle; a piece of beef
into rhomboids; and a pudding into a cycloid". Faced with such an unhealthy
diet of Euclidean shapes and inhabitants whose mathematical obsessions did not
extend to the world of nature Gulliver so loved, he was soon plotting to leave.

In this short passage, Swift was echoing the sentiments of countless math
students who, like Gulliver, when faced with a diet of Euclidean shapes failed
to see their relevance and turned to thoughts of escape. With the advent of
fractal geometry, however, nutrition has greatly improved. Broccoli and other
vegetables are now "acceptable" shapes and a modern-day Gulliver might never
leave the island of Laputa. As we have attempted to demonstrate in this article,
an obsession with mathematics need not turn away from nature. The Laputans
are now free to contemplate the beauty of the M-set or the geometry of their
own island and its surroundings (back cover).

1.6   Mathematical details and formalism

This section provides many of the formal mathematical details that were omitted
from the earlier sections.

1.6.1   Fractional Brownian motion

A fractional Brownian motion, $V_H(t)$, is a single valued function of one variable, t (usually time). Its increments $V_H(t_2) - V_H(t_1)$ have a Gaussian distribution with variance

$$<| V_H(t_2) - V_H(t_1) |^2> \propto | t_2 - t_1 |^{2H}, \qquad (1.24)$$

where the brackets $<$ and $>$ denote ensemble averages over many samples of
$V_H(t)$ and the parameter H has a value $0 < H < 1$. Such a function is both
stationary and isotropic. Its mean square increments depend only on the time
difference $t_2 - t_1$ and all t's are statistically equivalent. The special value $H = \frac{1}{2}$
gives the familiar Brownian motion with

$$\Delta V^2 \propto \Delta t.$$

As with the usual Brownian motion, although $V_H(t)$ is continuous, it is nowhere differentiable. Nevertheless, many constructs have been developed (and are relevant to the problem of light scattering from fractals) to give meaning to "derivative of fractional Brownian motion" as *fractional Gaussian noises* [63,68] of Figure 1.9 . Such constructs are usually based on averages of $V_H(t)$ over decreasing scales. The derivative of normal Brownian motion, $H = \frac{1}{2}$, corresponds to uncorrelated *Gaussian white noise* (Figure 1.9a) and Brownian motion is said to have *independent increments*. Formally, this implies that for any three times such that $t_1 < t < t_2$, $V_H(t) - V_H(t_1)$ is statistically independent of $V_H(t_2) - V_H(t)$ for $H = \frac{1}{2}$. For $H > \frac{1}{2}$ there is a positive correlation both for the increments of $V_H(t)$ and its derivative fractional Gaussian noise. For $H < \frac{1}{2}$ the increments are negatively correlated. Such correlations extend to arbitrarily long time scales and have a large effect on the visual appearance of the fBm traces as shown in Figure 1.12.

$V_H(t)$ shows a statistical scaling behavior. If the time scale t is changed by the factor r, then the increments ΔV_H change by a factor r^H. Formally,

$$< \Delta V_H(rt)^2 > \quad \propto \quad r^{2H} < \Delta V_H(t)^2 > . \qquad (1.25)$$

Unlike statistically self-similar curves (such as the coastlines in Figure 1.1), a $V_H(t)$ trace requires *different* scaling factors in the two coordinates (r for t but r^H for V_H) reflecting the special status of the t coordinate. Each t can correspond to only one value of V_H but any specific V_H may occur at multiple t's. Such non-uniform scaling is known as *self-affinity* rather than self-similarity.

1.6.2 Exact and statistical self-similarity

The distinction between similarity and affinity is important, but has only recently received wide attention [72]. By way of summary, a *self-similar* object is composed of N copies of itself (with possible translations and rotations) each of which is scaled down by the ratio r in all E coordinates from the whole. More formally, consider a set S of points at positions

$$\vec{x} = (x_1, \ldots, x_E)$$

in Euclidean space of dimension E. Under a *similarity* transform with real scaling ratio $0 < r < 1$, the set S becomes rS with points at

$$r\vec{x} = (rx_1, \ldots, rx_E)$$

A bounded set S is *self-similar* when S is the union of N distinct (non-overlapping) subsets each of which is congruent to rS. *Congruent* means identical

under translations and rotations. The fractal or *similarity dimension* of S is then
given by

$$1 = Nr^D \quad \text{or} \quad D = \frac{\log N}{\log \frac{1}{r}}. \qquad (1.26)$$

This relation leads to several important methods of *estimating D* for a given set
S as presented in Figure 1.16 and described below.

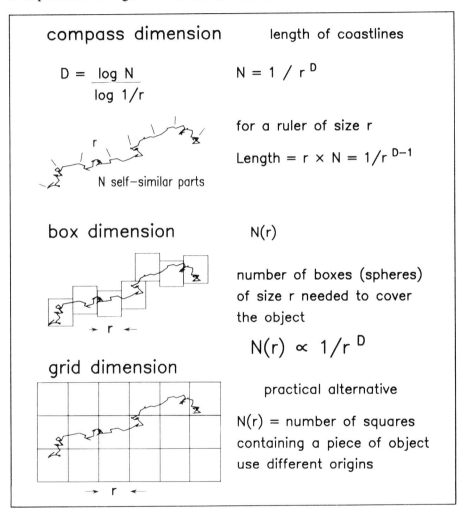

Fig. 1.16: Basic techniques for estimating fractal dimension from experimental images.

The set S is also *self-similar* if each of the N subsets is scaled down from
the whole by a different similarity ratio r_n. In this case, D is given implicitly
by

$$\sum_{n=1}^{N} r_n^D = 1, \qquad (1.27)$$

which reduces to the familiar result in Eqn. (1.26) when all of the r_n are equal.

The set S is *statistically self-similar* if it is composed of N distinct subsets each of which is scaled down by the ratio r from the original and is identical in all statistical respects to rS. The similarity dimension is again given by Eqn. (1.26). In practice, it is impossible to verify that all moments of the distributions are identical, and claims of statistical self-similarity are usually based on only a few moments. Moreover, a sample of a random set (such as a coastline) is often statistically self-similar for all scaling ratios r. Its fractal dimension is usually estimated from the dependence of box coverings $N_{box}(L)$ or mass $M(L)$ on varying L as in Eqs. (1.28) and (1.29).

1.6.3 Measuring the fractal dimension D

For topologically one-dimensional fractal "curves", the apparent "length" varies with the measuring ruler size. If the entire self-similar curve is of maximum size L_{max}, then, on a smaller scale $L = rL_{max}$ with $r < 1$, and the curve consists of $N = \frac{1}{r^D}$ segments of length L. Thus, for $D \geq 1$

$$\text{Length} = L \cdot N = L \cdot \left(\frac{L_{max}}{L}\right)^D \propto \frac{1}{L^{D-1}}. \qquad (1.28)$$

The fractal dimension D also characterizes the covering of the set S by E-dimensional "boxes" of linear size L. If the entire S is contained within one box of size L_{max}, then each of the $N = \frac{1}{r^D}$ subsets will fall within one box of size $L = rL_{max}$. Thus, the number of boxes of size L, $N_{box}(L)$, needed to cover S is given by

$$N_{box}(L) = \left(\frac{L_{max}}{L}\right)^D \quad \text{or} \quad N_{box}(L) \propto \frac{1}{L^D}. \qquad (1.29)$$

This definition of *box dimension* is one of the most useful methods for estimating the fractal dimension of a given set. The box dimension can be conveniently estimated by dividing the E-dimensional Euclidean space containing the set into a grid of boxes of size L^E and counting the number of such boxes $N_{box}(L)$ that are non-empty. It is useful to examine as large a range of L as possible and to average over various origins for the boxes.

One can also estimate the "volume" or "mass" of the set S by a covering with boxes of linear size L. If one considers only distances of order L about a given point in S, one finds a single box of size L with E-dimensional volume L^E. If the distance scale about the same point is increased to $L_{max} = \frac{L}{r}$, one now finds a total of

$$N = \frac{1}{r^D} = \left(\frac{L_{max}}{L}\right)^D$$

boxes of mass L^E covering the set. Thus, the mass within a distance L_{max} of some point in S,

$$M(L_{max}) = N \cdot M(L) = M(L) \cdot \left(\frac{L_{max}}{L}\right)^D$$

or

$$M(L) \propto L^D . \qquad (1.30)$$

The fractal dimension D, thus, also corresponds to the commonly used *mass dimension* in physics. Mass dimension also has a strong connection with intuitive notions of dimension. The amount of material within a distance L of a point in a one-dimensional object increases as L^1. For an E-dimensional object it varies as L^E. The mass dimension is another extremely useful method for estimating the fractal dimension of a given object.

For a distribution of objects in the plane, the fractal dimension of their coastlines can also be estimated from the perimeter vs area scaling of the islands. Eqn. (1.28) gives the *Length* or perimeter of a fractal object of size L_{max} when "measured" with a ruler of size L. Consider a population of such islands with the same coastline D and all measured with the same ruler L. For a given island, perimeter $P \propto L_{max}^D$ from Eqn. (1.28). Provided the coastline $D < 2$, each island will have a well-defined area $A \propto L_{max}^2$ and

$$P \propto A^{\frac{D}{2}}. \qquad (1.31)$$

A study of P vs A scaling is a useful method of estimating D for a population of fractal objects such as percolation clusters [104].

1.6.4   Self-affinity

Under an *affine* transform, on the other hand, each of the E coordinates of $\vec{x}$ may be scaled by a different ratio $(r_1, \ldots, r_E)$. Thus, the set S is transformed to $r(S)$ with points at $r(\vec{x}) = (r_1 x_1, \ldots, r_E x_E)$. A bounded set S is *self-affine* when S is the union of N distinct (non-overlapping) subsets each of which is congruent to $r(S)$. Similarly, S is *statistically self-affine* when S is the union of N distinct subsets each of which is congruent *in distribution* to $r(S)$. This is the case for the statistically self-affine fBm, $V_H(t)$, where the t- and V-coordinates must be scaled differently. The fractal dimension D, however, is not as easily defined as with self-similarity.

1.6.5 The relation of D to H for self-affine fractional Brownian motion

The assignment of a fractal dimension D (using the self-similar strategies presented above) to a self-affine set can produce ambiguous results [72]. The difficulties can be illustrated with the case of fractional Brownian motion $V_H(t)$. Consider, for convenience, a trace of $V_H(t)$ from Figure 1.12 covering a time span $\Delta t = 1$ and a vertical range $\Delta V_H = 1$. $V_H(t)$ is statistically self-affine when t is scaled by r and V_H is scaled by r^H. Suppose the time span is divided into N equal intervals each with $\Delta t = \frac{1}{N}$. Each of these intervals will contain one portion of $V_H(t)$ with vertical range

$$\Delta V_H = \Delta t^H = \frac{1}{N^H}.$$

Since $0 < H < 1$ each of these new sections will have a large vertical to horizontal size ratio and the occupied portion of each interval will be covered by

$$\Delta V_H \Delta t = \frac{\frac{1}{N^H}}{\frac{1}{N}} = \frac{N}{N^H}$$

square boxes of linear scale $L = \frac{1}{N}$. In terms of box dimension, as t is scaled down by a ratio $r = \frac{1}{N}$ the number of square boxes covering the trace goes from 1 to $N(L)$ = number of intervals times boxes per interval or

$$N(L) = N \cdot \frac{N}{N^H} = N^{2-H} = \frac{1}{L^{2-H}}.$$

Thus, by comparison with Eqn. (1.29),

$$D = 2 - H \quad \text{for a trace of} \quad V_H(t). \tag{1.32}$$

Consequently, the trace of normal Brownian motion is said to have $D = 1.5$.

It is important to note that the above association of a similarity dimension D with a self-affine fractal such as fBm is implicitly fixing a scaling between the (otherwise independent) coordinates. The result depends strongly on whether one is looking at scales large or small compared to this (artificially introduced) characteristic length and on which method chosen to estimate D. The difference is particularly clear when one attempts to estimate D for a trace of fBm from the "coastline" method of Eqn. (1.28). One can divide the curve into N segments by walking a ruler of size l along the curve. The length *along* each segment is

$$l = \sqrt{\Delta t^2 + \Delta V^2}$$

while the typical V variation scales with Δt as $\Delta V = \Delta t^H$. The number of rulers of size l, N(l) is expected to vary as $\frac{1}{l^D}$, where

$$l \propto \Delta t \sqrt{1 + \frac{\Delta V^2}{\Delta t^2}} \propto \Delta t \sqrt{1 + \frac{1}{\Delta t^{2-2H}}},$$

and $N = \frac{1}{\Delta t}$. On small scales with $\Delta t \ll 1$, the second term dominates and $l \propto \Delta t^H$ so $N(l) \propto l^{-\frac{1}{H}}$ and $D = \frac{1}{H}$. On the other hand, on large scales with $\Delta t \gg 1$, l varies linearly with Δt and $D = 1$. Thus, the same $V_H(t)$ trace can have an apparent self-similar dimension D of either 1, $\frac{1}{H}$, or $2 - H$ depending on the measurement technique and arbitrary choice of length scale.

1.6.6   Trails of fBm

Consider a particle undergoing a fractional Brownian motion or random walk in E dimensions where each of the coordinates is tracing out an independent fBm in time. Over an interval Δt each coordinate will vary by typically $\Delta L = \Delta t^H$. If overlap can be neglected, the "mass" of the trail

$$M \propto \Delta t \propto L^{\frac{1}{H}}.$$

In comparison with Eqn. (1.30), the trail of fBm is a self-similar curve with fractal dimension

$$D = \frac{1}{H} \qquad\qquad (1.33)$$

provided $\frac{1}{H} < E$. Normal Brownian motion with $H = 0.5$ has $D = 2$. $\frac{1}{H}$ is known as the *latent* fractal dimension [71,73] of a *trail* of fBm. When $\frac{1}{H} > E$ overlap cannot be neglected and the actual $D = E$. When $\frac{1}{H} = E$ the trail is *critical* and most quantities will have important logarithmic corrections to the usual fractal power laws. Note that although the *trail* of a fBm in E dimensions is self-similar, each of the E coordinates has a self-affine *trace* vs time.

1.6.7   Self-affinity in E dimensions

The formalism of a fBm vs time can be easily extended to provide a self-affine fractional Brownian function, V_H of $\vec{x} = (x_1, \dots, x_E)$ in E Euclidean dimensions. V_H satisfies the general scaling relation

$$< | V_H(\vec{x}_2) - V_H(\vec{x}_1) |^2 > \propto | \vec{x}_2 - \vec{x}_1 |^{2H} \qquad\qquad (1.34)$$

and has a fractal dimension

$$D = E + 1 - H. \qquad\qquad (1.35)$$

The zerosets of $V_H(\vec{x})$ form a statistically self-similar fractal with dimension $D_0 = E - H$.

1.6.8 Spectral densities for fBm and the spectral exponent β

Random functions in time $V(t)$ are often characterized [42,90] by their *spectral densities* $S_V(f)$. $S_V(f)$ gives information about the time correlations of $V(t)$. When $S_V(f)$ increases steeply at low f, $V(t)$ varies more slowly. If one defines $V(f,T)$ as the Fourier transform of a specific sample of $V(t)$ for $0 < t < T$,

$$V(f,T) = \frac{1}{T} \int_0^T V(t) e^{2\pi i f t} dt,$$

then

$$S_V(f) \propto T \mid V(f,T) \mid^2 \quad \text{as} \quad T \to \infty. \tag{1.36}$$

An alternate characterization of the time correlations of $V(t)$ is given by the *2 point autocorrelation function*

$$G_V(\tau) = <V(t)V(t+\tau)> - <V(t)>^2 . \tag{1.37}$$

$G_V(\tau)$ provides a measure of how the fluctuations at two times separated by τ are related. $G_V(\tau)$ and $S_V(f)$ are not independent. In many cases they are related by the Wiener-Khintchine relation [42,90,91]

$$G_V(\tau) = \int_0^\infty S_V(f) \cos(2\pi f \tau) df. \tag{1.38}$$

For a Gaussian white noise $S_V(f) = $ constant and $G_V(\tau) = \Delta V^2 \delta(\tau)$ is completely uncorrelated. For certain simple power laws for $S_V(f)$, $G_V(\tau)$ can be calculated exactly. Thus, for

$$S_V(f) \propto \frac{1}{f^\beta} \quad \text{with} \quad 0 < \beta < 1, \quad G_V(\tau) \propto \tau^{\beta-1}. \tag{1.39}$$

Moreover, $G_V(\tau)$ is directly related to the mean square increments of fBm,

$$<\mid V(t+\tau) - V(t) \mid^2> = 2 \left[<V^2> - G_V(\tau) \right]. \tag{1.40}$$

Roughly speaking, $S_V(f) \propto \frac{1}{f^\beta}$ corresponds to $G_V(\tau) \propto \tau^{1-\beta}$ and a fBm with $2H = \beta - 1$ from Eqs. (1.24) and (1.40). Thus, the statistically self-affine fractional Brownian function, $V_H(\vec{x})$, with $\vec{x}$ in an E-dimensional Euclidean space, has a fractal dimension D and spectral density $S_V(f) \propto \frac{1}{f^\beta}$, for the fluctuations along a straight line path in any direction in E-space with

$$D = E + 1 - H = E + \frac{3 - \beta}{2} \tag{1.41}$$

This result agrees with other "extensions" of the concepts of *spectral density* and *Wiener-Khintchine relation* to *non-stationary* noises where some moments may be undefined.

Although the formal definition of fBm restricts H to the range $0 < H < 1$, it is often useful to consider integration and an appropriate definition of "derivative" as extending the range of H. Thus, integration of a fBm produces a new fBm with H increased by 1, while "differentiation" reduces H by 1. When $H \to 1$, the derivative of fBm looks like a fBm with $H \to 0$. In terms of spectral density, if $V(t)$ has $S_V(f) \propto \frac{1}{f^\beta}$ then its derivative $\frac{dV}{dt}$ has spectral density $\frac{f^2}{f^\beta} = \frac{1}{f^{\beta-2}}$. In terms of Eqn. (1.41) , differentiation decreases β by 2 and decreases H by 1.

1.6.9   Measuring fractal dimensions: Mandelbrot measures

Numerical simulation or experimental image analysis often produces a geometrical object defined by a set S of points at positions $\vec{x} = (x_1, \ldots, x_E)$ in an E-dimensional Euclidean space. Lacking other information, all of the points may be assumed to be equivalent and all points are equally probable origins for analysis. The spatial arrangement of the points determines $P(m, L)$. $P(m, L)$ is the probability that there are m points within an E-cube (or sphere) of size L centered about an arbitrary point in S. $P(m, L)$ is normalized

$$\sum_{m=1}^{N} P(m, L) = 1 \quad \text{for all} \quad L. \tag{1.42}$$

$P(m, L)$ is directly related to other probability measures as used by Mandelbrot [65,68,71] Hentschel and Procaccia [52], and others [85]. This particular definition of $P(m, L)$ is, however, particularly efficient to implement on a computer.

The usual quantities of interest are derived from the moments of $P(m, L)$. The *mass dimension,*

$$M(L) = \sum_{m=1}^{N} m P(m, L). \tag{1.43}$$

The number of boxes of size L needed to cover S,

$$N_{box}(L) = \sum_{m=1}^{N} \frac{1}{m} P(m, L). \tag{1.44}$$

The configurational entropy when space is divided into cubes of size L

$$S(L) = \sum_{m=1}^{N} \log m P(m, L). \tag{1.45}$$

For a fractal set $M(L) \propto L^D$, $N_{box}(L) \propto \frac{1}{L^D}$, and $e^{S(L)} \propto L^D$. In fact, one can define all moments

$$M^q(L) = \sum_{m=1}^{N} m^q P(m, L),$$ (1.46)

for $q \neq 0$ (the $q = 0$ case is given by Eqn. (1.45) above) and one can then estimate D from the logarithmic derivatives

$$D_q = \frac{1}{q} < \frac{\partial \log M^q(L)}{\partial \log L} > \quad \text{for} \quad q \neq 0,$$ (1.47)

and

$$D = \frac{1}{q} < \frac{\partial S(L)}{\partial \log L} > \quad \text{for} \quad q = 0.$$ (1.48)

A double logarithmic plot of $M^q(L)$ vs L is an essential tool in verifying whether a fractal interpretation is valid for S. One expects a fractal to have power-law behavior of $M^q(L)$ over a wide range of L.

For a uniform fractal (fractal set) as the number of points examined, $N \to \infty$ the distribution is expected to take the scaling form $P(m, L) \to f(\frac{m}{L^D})$ and all moments give the same D. For a non-uniform fractal (a fractal measure such as a Poincare map) the moments may take different values. Shapes and measures requiring more than one D are known as *multi-fractals*. They have received wide attention in studying the properties of dynamical systems (Poincare maps) and growth probability (or electric field strength) in aggregation problems. Different methods of characterizing the distribution of D's have been used to compare theory and experiment.

Figure 1.17 demonstrates the use of Eqs. (1.46) and (1.47) in estimating D_q. Figure 1.17(a) shows the measured dependence of $< M^q(L) >^{\frac{1}{q}}$ on L for several q for the exactly self-similar triadic Koch curve. In this case, as expected, all of the moments show the same dependence on L and the estimates of D_q from the logarithmic slopes agree well with the exact value of $\frac{\log(4)}{\log(3)} = 1.2618....$ Figure 1.17(b) shows the same estimates for the computer generated coastline of Figure 1.1. Here slight differences between the various moments are visible with D_q increasing slightly with q.

1.6.10 Lacunarity

It is obvious from the above discussion that the fractal dimension D characterizes only part of the information in the distribution $P(m, L)$. Different fractal sets may share the same D but have different appearances or *textures* corresponding to different $P(m, L)$. As an initial step toward quantifying texture,

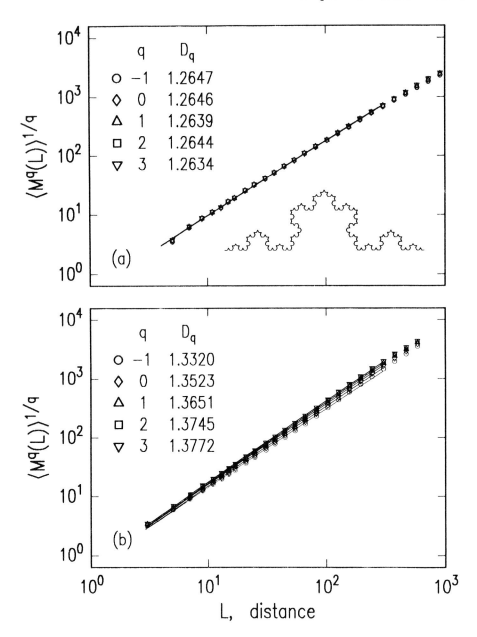

Fig. 1.17: Use of multi-fractal $M^q(L)$ in experimental analysis. **a.** Von Koch curve. **b.** Coastline from Figure 1.1 .

Mandelbrot ([68], Chapter 34) has introduced the parameter *lacunarity* , Λ (*lacuna* is Latin for gap). Although the qualitative visual effect of changing lacunarity at fixed D is quite striking as shown in Figure 1.15, to date there have been no quantitative measurements of the lacunarity of various random fractals. Mandelbrot [68] offers several alternative definitions. One derives from the width of the distribution $P(m, L)$ at fixed L. Given $M^q(L)$ as defined by

Eqn. (1.46),

$$\Lambda(L) = \frac{<M^2(L)> - <M(L)>^2}{<M(L)>^2}.\qquad(1.49)$$

When $P(m, L)$ takes the scaling form $f(\frac{m}{L^D})$, Λ is just the relative mean square width of the distribution f,

$$\Lambda = \frac{<\Delta^2>}{<f>^2}.$$

Figure 1.15 shows fractal surfaces of the same D with different lacunarity.

1.6.11   Random cuts with $H \neq \frac{1}{2}$: Campbell's theorem

The equivalence of Brownian motion to a summation of step functions is a special case of Campbell's theorem (1909). Consider a collection of independent pulses occurring at random times t_i corresponding to an average rate $\frac{1}{\tau}$. Each pulse produces the profile $A_i P(t - t_i)$. The random function $V(t) = \sum A_i P(t - t_i)$, will then have a spectral density [42],

$$S_V(f) \propto \frac{1}{\tau} <A^2> |P(f)|^2,\qquad(1.50)$$

where $P(f)$ is the Fourier transform of $P(t)$, $P(f) = \int P(t)e^{2\pi i ft}dt$. For normal Brownian motion, each excitation by the white noise $W(t)$ produces the step function response $P(t)$ with $P(f) \propto \frac{1}{f}$ and $S_V(f) \propto \frac{1}{f^2}$. With a suitable power-law choice of $P(t)$, one can generate an approximation to fBm with any H. In fact, fBm can be put in a form [63,68] similar to normal Brownian motion

$$V_H(t) = \frac{1}{\Gamma\left(H + \frac{1}{2}\right)} \int_{-\infty}^{t} (t - s)^{H-\frac{1}{2}} W(s)\, ds.\qquad(1.51)$$

Thus, fBm can be constructed from independent pulses with response $P(t) = t^{H-\frac{1}{2}}$ corresponding to $P(f) \propto \frac{1}{f^{H+\frac{1}{2}}}$ from Eqn. (1.36) and $S_V(f) \propto \frac{1}{f^{2H+1}}$ in agreement with Eqn. (1.39).

1.6.12   FFT filtering in 2 and 3 dimensions

In Section 1.3 the fractional Brownian surface was defined as having the same correlations in the plane as a fractional Brownian motion of the same H has in time. Thus, $V_H(x, y)$ should have the same form for its autocorrelation function as a fBm. Eqn. (1.38) can be extended to the xy plane

$$<V_H(\vec{r})V_H(\vec{r}+\vec{\delta})> = \int_0^\infty S(\vec{k})\cos(2\pi\vec{k}\cdot\vec{\delta})2\pi k\, dk,\qquad(1.52)$$

where $\vec{r}$ is a point in the xy plane and $\vec{\delta}$ is a displacement in the xy plane. Since all directions in the xy plane are equivalent, $S(\vec{K})$ depends only on

$$|\vec{k}| = k = \sqrt{k_x^2 + k_y^2}$$

and $|\vec{\delta}|$. For this δ dependence to correspond to the τ dependence of $<V(t)V(t+\tau)>$, $S(\vec{k})$ must vary as $\frac{1}{k^{1+\beta}}$ and $S(\vec{k}) \propto \frac{S_{cut}(k)}{k}$. The extra factor of k compensates for the 2 dimensional differential, $2\pi k dk$, in the integrand. For 3 coordinates, the spectral density of $V_H(x, y, z)$, $S(\vec{k}) \propto \frac{S_{cut}(k)}{k^2}$ and $S(\vec{k})$ varying as $\frac{1}{k^{2+\beta}}$ produces a $\frac{1}{f^\beta}$-noise for a sample V_H along any line.

With $E = 2$ the corresponding fractal surface is approximated on a finite N by N grid to give $V_H(x_n, y_m)$ where $x_n = n\lambda$ and $y_m = m\lambda$. The 2 dimensional complex FFT can be used to evaluate the series

$$V_{nm} = \sum_{qr}^{\frac{N}{2}-1} v_{qr} e^{2\pi i(k_q x_n + k_r y_m)}, \tag{1.53}$$

where $k_q = \frac{q}{N\lambda}$ and $k_r = \frac{r}{N\lambda}$ are the spatial frequencies in the x and y directions. For a fBm surface corresponding to $\beta = 1 + 2H = 7 - 2D$, the coefficients v_{qr} must satisfy

$$<|v_{qr}|^2> \propto \frac{1}{k^{\beta+1}} \propto \frac{1}{(q^2 + r^2)^{4-D}}. \tag{1.54}$$

The landscapes of Plates 8 to 13 with $2 < D < 3$ were generated with this process.

For a 3-dimensional fractional Brownian volume, $V_H(x, y, z)$, the differential in the autocorrelation function contains k^2 and the Fourier coefficients v_{qrs} will satisfy

$$<|v_{qrs}|^2> \propto \frac{1}{k^{\beta+2}} \propto \frac{1}{\left(\sqrt{q^2 + r^2 + s^2}\right)^{11-2D}} \tag{1.55}$$

with $3 < D < 4$.

Chapter 2

Algorithms for random fractals

Dietmar Saupe

2.1 Introduction

For about 200 years now mathematicians have developed the theory of smooth curves and surfaces in two, three or higher dimensions. These are curves and surfaces that globally may have a very complicated structure but in small neighborhoods they are just straight lines or planes. The discipline that deals with these objects is differential geometry. It is one of the most evolved and fascinating subjects in mathematics. On the other hand fractals feature just the opposite of smoothness. While the smooth objects do not yield any more detail on smaller scales a fractal possesses infinite detail at all scales no matter how small they are. The fascination that surrounds fractals has two roots: Fractals are very suitable to simulate many natural phenomena. Stunning pictures have already been produced, and it will not take very long until an uninitiated observer will no longer be able to tell whether a given scene is natural or just computer simulated. The other reason is that fractals are simple to generate on computers. In order to generate a fractal one does not have to be an expert of an involved theory such as calculus, which is necessary for differential geometry. More importantly, the complexity of a fractal, when measured in terms of the length of the shortest computer program that can generate it, is very small.

71

Fractals come in two major variations. There are those that are composed of several scaled down and rotated copies of themselves such as the von Koch snowflake curve or the Sierpinski gaskets. Julia sets also fall into this general category, in the sense that the whole set can be obtained by applying a nonlinear iterated map to an arbitrarily small section of it (see Chapter 3). Thus, the structure of Julia sets is already contained in any small fraction. All these fractals can be termed *deterministic fractals*. Their computer graphical generation requires use of a particular mapping or rule which then is repeated over and over in a usually recursive scheme.

In the fractals discussed here the additional element of randomness is included, allowing simulation of natural phenomena. Thus we call these fractals *random fractals*.

Given that fractals have infinite detail at all scales it follows that a complete computation of a fractal is impossible. So approximations of fractals down to some finite precision have to suffice. The desired level of resolution is naturally given by constraints such as the numbers of pixels of the available graphics display or the amount of compute-time that one is willing to spend. The algorithms presented below fall into several categories.

(C1) An approximation of a random fractal with some resolution is used as input and the algorithm produces an improved approximation with resolution increased by a certain factor. This process is repeated with outputs used as new inputs until the desired resolution is achieved. In some cases the procedure can be formulated as a recursion. The *midpoint displacement methods* are examples.

(C2) Only one approximation of a random fractal is computed namely for the final resolution. Pictures for different resolutions are possible but require that most of the computations have to be redone. The *Fourier filtering method* is an example.

(C3) In the third approach the approximation of a fractal is obtained via an iteration. After each step the approximation is somewhat improved, but the spatial resolution does not necessarily increase by a constant factor in each iteration. Here the allowed compute-time determines the quality of the result. The *random cut method* described below or the computation of Julia sets as preimages of repellers are examples.

Almost all of the algorithms contained in this paper are based on methods and ideas discussed in Chapter 1 as well as in [103] and ultimately in [68], where also some background information and further references can be found. Please note also the following two disclaimers:

It is not the intent of this chapter to give a comprehensive account of the theory that leads to the algorithms. Instead we have to direct the interested reader to Chapter 1 and the references therein. Secondly, we describe only a few of the aspects of the rendering of fractals. It must be acknowledged that the generation of data of a fractal alone is not sufficient to obtain a good picture. It is likely that computer programmers will spend far more time writing software that can render e.g. a fractal landscape with some reasonable coloring and shading (if that work has not already been done) than for the coding of fractal data generation routines. In any case the computer will surely use much more CPU time for rendering than for data generation. This is a good reason not to be satisfied with poor approximations of fractals due to insufficient algorithms.

In the following sections we describe a number of algorithms with increasing complexity. They are documented in the form of pseudo code which we hope is self-explanatory. This code is included to clarify the methods and therefore is not optimized for speed. Also, it is not complete. Some routines such as those for fast Fourier transformation and multilinear interpolation have been omitted. If they are not available in the form of library routines, they can easily be written using the indicated literature.

In Section 2.7 we explain two methods to graphically display fractal surfaces:

a. flat two-dimensional top view and parallel projection for three-dimensional view using color mapped elevation, Gouraud shaded polygons, and the painters algorithm for hidden surface elimination.

b. three-dimensional perspective representation with points as primitives, using an extended floating horizon method.

In Section 2.8 we list a number of definitions from probability theory and related to random processes. Notions such as "random variables", "correlation" etc. are marked with a dagger (†) where they first appear in the main text, and they are then briefly explained in Section 2.8 for the convenience of those readers who would like to remind themselves about these terms.

2.2   First case study: One-dimensional Brownian motion

2.2.1   Definitions

Brownian motion in one variable constitutes the simplest random fractal, and also it is at the heart of all the following generalizations. Small particles of solid matter suspended in a liquid can be seen under a microscope to move about in an irregular and erratic way. This was observed by the botanist R. Brown around 1827. The modeling of this movement is one of the great topics of statistical mechanics (see e.g. [90]). Occasionally Brownian motion is also referred to as "brown noise". In one dimension we obtain a random process† $X(t)$, i.e. a function X of a real variable t (time) whose values are random variables† $X(t_1), X(t_2)$, etc.

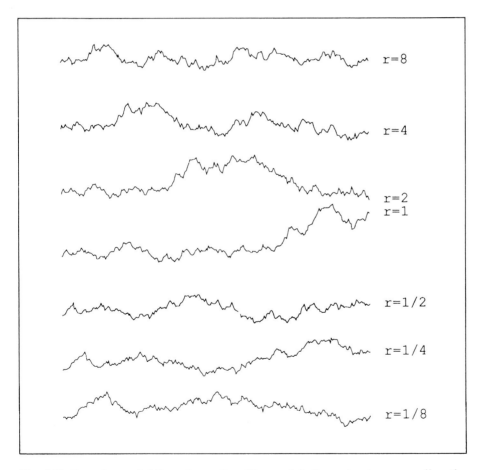

Fig. 2.1: Properly rescaled Brownian motion. The graph in the center shows a small section of Brownian motion $X(t)$. In the other graphs the properly rescaled random functions of the form $r^{-\frac{1}{2}}X(rt)$ are displayed. The scaling factor r ranges from $r = \frac{1}{8}$ to $r = 8$ corresponding to expanding and contracting the original function in the time direction. Note, that the visual appearance of all samples is the same.

The increment

$$X(t_2) - X(t_1) \text{ has Gaussian distribution}\dagger \qquad (2.1)$$

and the mean square increments† have a variance proportional to the time differences, thus

$$E\left[|X(t_2) - X(t_1)|^2\right] \propto |t_2 - t_1|. \qquad (2.2)$$

Here E denotes the mathematical expectation† of a random variable, or, in other words, the average over many samples. We say that the increments of X are *statistically self-similar* in the sense that

$$X(t_0 + t) - X(t_0) \quad \text{and} \quad \frac{1}{\sqrt{r}}\left(X(t_0 + rt) - X(t_0)\right)$$

have the same finite dimensional joint distribution† functions for any t_0 and $r > 0$. If we take for convenience $t_0 = 0$ and $X(t_0) = 0$ then this means that the two random functions

$$X(t) \quad \text{and} \quad \frac{1}{\sqrt{r}}X(rt)$$

are statistically indistinguishable. The second one is just a *properly rescaled* version of the first. Thus, if we accelerate the process $X(t)$ by a factor of 16, for example, then we can divide $X(16t)$ by 4 to obtain the same Brownian motion that we started with. We will return to this important characterization, when we discuss fractional Brownian motion, and it will be crucial in understanding the spectral synthesis method.

2.2.2 Integrating white noise

The integral of uncorrelated white Gaussian noise W satisfies (2.1) and (2.2):

$$X(t) = \int_{-\infty}^{t} W(s)\,ds. \qquad (2.3)$$

The random variables $W(t)$ are uncorrelated† and have the same normal distribution† $N(0,1)$. Moreover, the graph of a sample of Brownian motion $X(t)$ has a fractal dimension of 1.5, and the intersection of the graph with a horizontal line has a dimension of 0.5 . Formula (2.3) gives rise to the first little algorithm *WhiteNoiseBM()*.

ALGORITHM **WhiteNoiseBM** (X, N, seed)		
Title	Brownian motion by integration of white Gaussian noise	
Arguments	X[]	array of reals of size N
	N	size of array X
	seed	seed value for random number generator
Variables	i	integer

```
BEGIN
    X[0] := 0
    InitGauss (seed)
    FOR i := 1 TO N-1 DO
        X[i] := X[i-1] + Gauss () / (N-1)
    END FOR
END
```

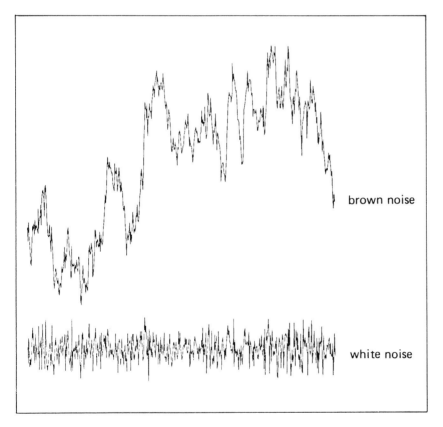

Fig. 2.2: Brownian motion in one dimension. The lower graph shows a sample of Gaussian white noise, its time integral yields the brown noise above it (see algorithm *WhiteNoiseBM()*).

2.2.3   Generating Gaussian random numbers

In the above algorithm we perform a simple numerical integration of white Gaussian noise. It requires a routine called *Gauss()*, which returns a sample

ALGORITHM **InitGauss** (seed)		
Title	Initialization of random number generators	

Arguments	seed	seed value for random number generator
Globals	Arand	rand() returns values between 0 and Arand, system dependent
	Nrand	number of samples of rand() to be taken in Gauss()
	GaussAdd	real parameter for the linear transformation in Gauss()
	GaussFac	real parameter for the linear transformation in Gauss()
Functions	srand()	initialization of system random numbers

```
BEGIN
    Nrand := 4
    Arand := power (2, 31) - 1
    GaussAdd := sqrt (3 * Nrand)
    GaussFac := 2 * GaussAdd / (Nrand * Arand)
    srand (seed)
END
```

ALGORITHM **Gauss** ()		
Title	Function returning Gaussian random number	

Globals	Nrand	number of samples of rand() to be taken in Gauss()
	GaussAdd	real parameter for the linear transformation in Gauss()
	GaussFac	real parameter for the linear transformation in Gauss()
Locals	sum	real
	i	integer
Functions	rand()	system function for random numbers

```
BEGIN
    sum := 0
    FOR i := 1 TO Nrand DO
        sum := sum + rand ()
    END FOR
    RETURN (GaussFac * sum - GaussAdd)
END
```

of a random variable with normal distribution (mean 0 and variance 1). Let us briefly describe an elementary method for generating such numbers. On most machines a pseudo random number generator will be available. So let us assume that we have a routine *rand()*, which returns random numbers uniformly distributed† over some interval $[0, A]$. Typically A will be $2^{31} - 1$ or $2^{15} - 1$. Also we assume that a routine *srand(seed)* exists, which introduces a seed value for *rand()*. Taking certain linearly scaled averages of the values returned by *rand()* will approximate a Gaussian random variable as follows:

A random variable Y is standardized by subtracting its expected value and dividing by its standard deviation†:

$$Z = \frac{Y - E(Y)}{\sqrt{\text{var } Y}}.$$

The Central Limit Theorem states, that if Z_n is the standardized sum of any n identically distributed random variables, then the probability distribution† of Z_n tends to the normal distribution as $n \to \infty$. Let Y_i be the i-th value returned by *rand()*, $i = 1, \ldots, n$. Then

$$E(Y_i) = \frac{1}{2}A, \ \text{var} \ Y_i = \frac{1}{12}A^2$$

and thus

$$E\left(\sum_{i=1}^{n} Y_i\right) = \frac{n}{2}A, \ \text{var}\left(\sum_{i=1}^{n} Y_i\right) = \frac{n}{12}A^2 .$$

With these formulas we obtain

$$Z_n = \frac{\sum_{i=1}^{n} Y_i - \frac{n}{2}A}{\sqrt{\frac{n}{12}}A} = \frac{1}{A}\sqrt{\frac{12}{n}}\sum_{i=1}^{n} Y_i - \sqrt{3n}$$

as our approximate Gaussian random variable. In practice $n = 3$ or 4 already yields satisfactory results for our purposes.

Let us remark that there are other Gaussian random number generators that require only one evaluation of *rand()* for each Gaussian random number returned. They use a transformation method, and the underlying distribution is exactly the normal distribution. We refer the interested reader to [8] and [86].

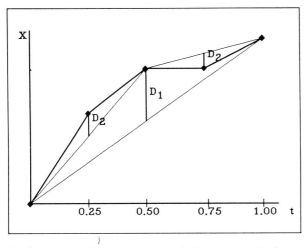

Fig. 2.3: Midpoint displacement.The first two stages in the midpoint displacement technique as explained in the text.

2.2.4   Random midpoint displacement method

Another straightforward method to produce Brownian motion is random midpoint displacement. If the process is to be computed for times, t, between 0 and

```
ALGORITHM MidPointBM  (X, maxlevel, sigma, seed)
Title            Brownian motion via midpoint displacement
```

Arguments	X[]	real array of size $2^{maxlevel} + 1$
	maxlevel	maximal number of recursions
	sigma	initial standard deviation
	seed	seed value for random number generator
Globals	delta[]	array holding standard deviations Δ_i
Variables	i, N	integers

```
BEGIN
    InitGauss (seed)
    FOR i := 1 TO maxlevel DO
        delta[i] := sigma * power (0.5, (i+1)/2)
    END FOR
    N = power (2, maxlevel)
    X[0] := 0
    X[N] := sigma * Gauss ()
    MidPointRecursion (X, 0, N, 1, maxlevel)
END
```

```
ALGORITHM MidPointRecursion  (X, index0, index2, level, maxlevel)
Title            Recursive routine called by MidPointBM()
```

Arguments	X[]	real array of size $2^{maxlevel} + 1$
	index0	lower index in array
	index2	upper index in array
	level	depth of the recursion
	maxlevel	maximal number of recursions
Globals	delta	array holding standard deviations Δ_i
Variables	index1	midpoint index in array

```
BEGIN
    index1 := (index0 + index2) / 2
    X[index1] := 0.5 * (X[index0] + X[index2]) + delta[level] * Gauss ()
    IF (level < maxlevel) THEN
        MidPointRecurs (X, index0, index1, level+1, maxlevel)
        MidPointRecurs (X, index1, index2, level+1, maxlevel)
    END IF
END
```

1, then one starts by setting $X(0) = 0$ and selecting $X(1)$ as a sample of a Gaussian random variable with mean 0 and variance σ^2. Then $\mathrm{var}(X(1) - X(0)) = \sigma^2$ and we expect

$$\mathrm{var}(X(t_2) - X(t_1)) = |t_2 - t_1|\sigma^2 \qquad (2.4)$$

for $0 \leq t_1 \leq t_2 \leq 1$. We set $X(\frac{1}{2})$ to be the average of $X(0)$ and $X(1)$ plus some Gaussian random offset D_1 with mean 0 and variance Δ_1^2. Then

$$X(\frac{1}{2}) - X(0) = \frac{1}{2}(X(1) - X(0)) + D_1$$

and thus $X(\frac{1}{2}) - X(0)$ has mean value 0 and the same holds for $X(1) - X(\frac{1}{2})$. Secondly, for (2.4) to be true we must require

$$\text{var}(X(\frac{1}{2}) - X(0)) = \frac{1}{4}\text{var}(X(1) - X(0)) + \Delta_1^2 = \frac{1}{2}\sigma^2.$$

Therefore

$$\Delta_1^2 = \frac{1}{4}\sigma^2.$$

In the next step we proceed in the same fashion setting

$$X(\frac{1}{4}) - X(0) = \frac{1}{2}(X(0) + X(\frac{1}{2})) + D_2$$

and observe that again the increments in X, here $X(\frac{1}{2}) - X(\frac{1}{4})$ and $X(\frac{1}{4}) - X(0)$ are Gaussian and have mean 0. So we must choose the variance Δ_2^2 of D_2 such that

$$\text{var}(X(\frac{1}{4}) - X(0)) = \frac{1}{4}\text{var}(X(\frac{1}{2}) - X(0)) + \Delta_2^2 = \frac{1}{4}\sigma^2$$

holds, i.e.

$$\Delta_2^2 = \frac{1}{8}\sigma^2.$$

We apply the same idea to $X(\frac{3}{4})$ and continue to finer resolutions yielding

$$\Delta_n^2 = \frac{1}{2^{n+1}}\sigma^2$$

as the variance of the displacement D_n. Thus corresponding to time differences $\Delta t = 2^{-n}$ we add a random element of variance $2^{-(n+1)}\sigma^2$ which is proportional to Δt as expected.

2.2.5 Independent jumps

Our last algorithm for Brownian motion falls into the category (C3). We may interpret Brownian motion as the cumulative displacement of a series of independent jumps, i.e. an infinite sum of functions

$$J_i(t) = A_i\beta(t - t_i)$$

where

$$\beta(t) = \begin{cases} 0, & \text{if } t < 0 \\ 1, & \text{if } t \geq 0 \end{cases}$$

and A_i, t_i are random variables with Gaussian and Poisson distributions respectively. This approach generalizes to circles and spheres. For circles we take time as 2π-periodic and arrive at

$$X(t) = \sum_{i=0}^{\infty} A_i \overline{\beta}(t - t_i), \qquad (2.5)$$

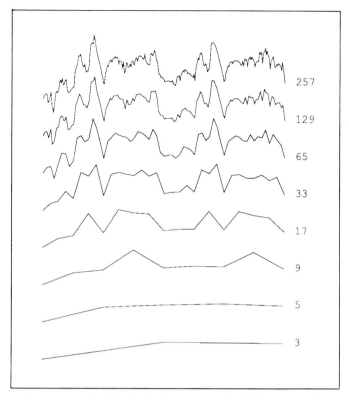

257

129

65

33

17

9

5

3

Fig. 2.4: Brownian motion via midpoint displacement method. Eight intermediate stages of the algorithm *MidPointBM( )* are shown depicting approximations of Brownian motion using up to 3, 5, 9, ..., 257 points, i.e. the parameter maxlevel was set to 2, 3, ..., 8.

where

$$\overline{\beta}(t) = \begin{cases} 0, \text{if } t \geq \pi(\bmod 2\,\pi) \\ 1, \text{if } t < \pi(\bmod 2\,\pi) \end{cases}$$

The A_i are identically distributed Gaussian random variables and the t_i are uniformly distributed random variables with values between 0 and 2π. Each term in the sum (2.5) adds random displacement on half of the circle.

Of course, we can also use midpoint displacement to obtain Brownian motion on a circle. We would just have to require that $X(0) = X(1)$ in order to maintain continuity. However, the midpoint method does not generalize to spheres whereas the random cut method does: In each step a great circle of the sphere is picked at random and one of the two half spheres is displaced by an amount determined by a Gaussian random variable. The pictures of the planets in Figure 1.2 and the Plates 6, 7 and 36 were produced in this way (see also [103,68] and Sections 0.3, 1.1.8).

```
ALGORITHM RandomCutsBM (X, N, maxsteps, seed)
Title           Brownian motion on a circle by random cuts

Arguments    X[]          array of reals of size N
             N            size of array X
             maxsteps     maximal number of displacements
             seed         seed value for random number generator
Globals      Arand        rand() returns values between 0 and Arand
Variables    k,k0,k1,step  integers

BEGIN
    InitGauss (seed)
    FOR k := 0 TO N-1 DO
        X[k] := 0
    END FOR
    FOR step := 1 TO maxsteps DO
        k0 := N * rand () / Arand
        k1 := k0 + N/2 - 1
        FOR k := k0 TO k1 DO
            IF (k < N) THEN
                X[i] := X[i] + Gauss()
            ELSE
                X[k-N] := X[k-N] + Gauss()
            END IF
        END FOR
    END FOR
END
```

2.3 Fractional Brownian motion : Approximation by spatial methods

2.3.1 Definitions

In the last section we studied random processes $X(t)$ with Gaussian increments and

$$\text{var}(X(t_2) - X(t_1)) \propto |t_2 - t_1|^{2H} \qquad (2.6)$$

where $H = \frac{1}{2}$. The generalization to parameters $0 < H < 1$ is called *fractional Brownian motion* (fBm, see [63], [68]). As in the case of ordinary Brownian motion, we say that the increments of X are *statistically self-similar with parameter H*, in other words

$$X(t_0 + t) - X(t_0) \text{ and } \frac{1}{r^H}(X(t_0 + rt) - X(t_0))$$

have the same finite dimensional joint distribution functions for any t_0 and $r > 0$. If we again use for convenience $t_0 = 0$ and $X(t_0) = 0$, the two random functions

$$X(t) \text{ and } \frac{1}{r^H}X(rt) \qquad (2.7)$$

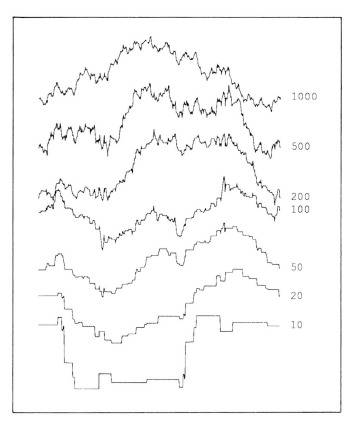

Fig. 2.5: Brownian motion via random cuts. Up to 1000 random cuts have been performed by the algorithm *RandomCutsBM( )*. All approximations are step functions which is clearly visible in the lower graphs.

are statistically indistinguishable. Thus "accelerated" fractional Brownian motion $X(rt)$ is *properly rescaled* by dividing amplitudes by r^H.

Let us visualize this important fact (see Figures 2.1 and 2.6). If we set $H = \frac{1}{2}$ we obtain the usual Brownian motion of the last section. For $H = 0$ we get a completely different behavior of X: We can expand or contract the graph of X in the t-direction by any factor, and the process will still "look" the same. This clearly says that the graph of a sample of X must densely fill up a region in the plane. In other words, its fractal dimension is 2. The opposite case is given by the parameter $H = 1$. There we must compensate for an expansion of the graph in the t-direction by also multiplying the amplitudes by the same factor. It is easy to give an argument showing that the fractal dimension of this graph must be $2 - H = 1$. In fact, graphs of samples of fBm have a fractal dimension of $2 - H$ for $0 < H < 1$. The parameter H thus describes the "roughness" of the function at small scales.

Fractional Brownian motion can be divided into three quite distinct cate-

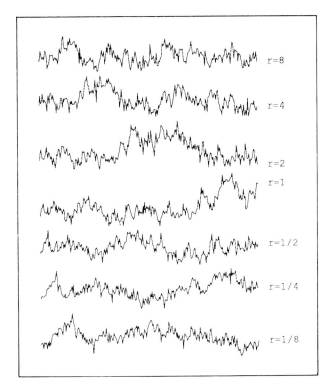 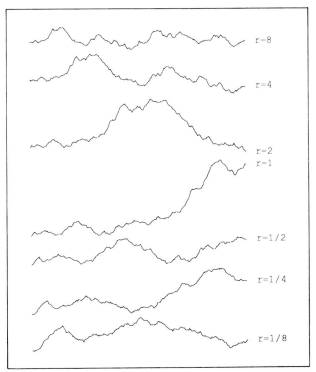

Fig. 2.6: Properly rescaled fractional Brownian motion. The two sets of curves show properly rescaled fractional Brownian motion for parameters $H = 0.2$ (left) and $H = 0.8$ (right). The graphs in the center show small sections of fractional Brownian motion $X(t)$. In the other graphs the properly rescaled random functions $r^{-H}X(rt)$ are displayed. The scaling factor r ranges from $r = \frac{1}{8}$ to $r = 8$ corresponding to expansion and contraction the original function in the time direction. Compare Figure 2.1 for $H = 0.5$.

gories: $H < \frac{1}{2}$, $H = \frac{1}{2}$ and $H > \frac{1}{2}$. The case $H = \frac{1}{2}$ is the ordinary Brownian motion which has independent increments, i.e. $X(t_2) - X(t_1)$ and $X(t_3) - X(t_2)$ with $t_1 < t_2 < t_3$ are independent in the sense of probability theory, their correlation† is 0. For $H > \frac{1}{2}$ there is a positive correlation between these increments, i.e. if the graph of X is increasing for some t_0, then it tends to continue to increase for $t > t_0$. For $H < \frac{1}{2}$ the opposite is true. There is a negative correlation of increments, and the curves seem to oscillate more erratically.

2.3.2 Midpoint displacement methods

For the approximation of fBm the approach taken by the midpoint method can formally be extended to also suit parameters $H \neq \frac{1}{2}$. Here we aim for the equivalent of (2.4)

$$\text{var}(X(t_2) - X(t_1)) = |t_2 - t_1|^{2H}\sigma^2.$$

Using the same line of thought as in the last section, we arrive at midpoint displacements D_n that have variances

$$\Delta_n^2 = \frac{\sigma^2}{(2^n)^{2H}}(1 - 2^{2H-2}). \qquad (2.8)$$

Therefore the only changes which are required in the algorithm *MidPointBM()* occur in the computation of Δ_n accommodating $H \neq \frac{1}{2}$.

ALGORITHM **MidPointFM1D** (X, maxlevel, sigma, H, seed)		
Title	One-dimensional fractal motion via midpoint displacement	

Arguments	X[]	real array of size $2^{maxlevel} + 1$
	maxlevel	maximal number of recursions
	sigma	initial standard deviation
	H	$0 < H < 1$ determines fractal dimension $D = 2 - H$
	seed	seed value for random number generator
Globals	delta[]	array holding standard deviations Δ_i
Variables	i, N	integers

```
BEGIN
    InitGauss (seed)
    FOR i := 1 TO maxlevel DO
        delta[i] := sigma * power (0.5, i*H) * sqrt (1 - power (2, 2*H-2))
    END FOR
    N = power (2, maxlevel)
    X[0] := 0
    X[N] := sigma * Gauss ()
    MidPointRecursion (X, 0, N, 1, maxlevel)
END
```

It has been shown that the above midpoint displacement technique does not yield true fBm for $H \neq \frac{1}{2}$ [69]. In fact, although

$$\text{var}(X(\frac{1}{2}) - X(0)) = \text{var}(X(1) - X(\frac{1}{2})) = (\frac{1}{2})^{2H}\sigma^2,$$

we do not have

$$\text{var}(X(\frac{3}{4}) - X(\frac{1}{4})) = (\frac{1}{2})^{2H}\sigma^2,$$

as we would like. Thus, this process does not have stationary increments, the times t are not all statistically equivalent. This defect causes the graphs of X to show some visible traces of the first few stages in the recursion. In the two-dimensional extension of this algorithm one obtains results which may exhibit some disturbing creases along straight lines, which are related to the underlying grid of points (compare Section 1.4.3 and the discussion in Appendix A). Nevertheless, this is still a useful algorithm for many purposes. It became most popular after its appearance in [40] and subsequently in some science magazines. Recently it has also been included as a part of some computer graphics text books [49],[50].

One approach to deal with the non-stationarity of the midpoint displacement technique is to interpolate the midpoints in the same way, but then to add a displacement D_n of a suitable variance to all of the points and not just the midpoints. This seems natural, as one would reiterate a measurement at all points in a graph of fBm, when a device is used that allows measurements at a smaller sampling rate Δt and a smaller spatial resolution ΔX. This method is called *successive random addition*. The extra amount of work involved as compared with the midpoint method is tolerable, about twice as many displacements are necessary. The actual formula for the variances of the displacements as used in the various stages of this algorithm (see pseudo code) will be derived as a special case of the method dicussed in the next section.

```
ALGORITHM AdditionsFM1D  (X, maxlevel, sigma, H, seed)
Title           One-dimensional fractal motion via successive random additions
```

Arguments	X[]	real array of size $2^{maxlevel} + 1$
	maxlevel	maximal number of recursions
	sigma	initial standard deviation
	H	$0 < H < 1$ determines fractal dimension $D = 2 - H$
	seed	seed value for random number generator
Globals	delta	array holding standard deviations Δ_i
Variables	i, N, d, D	integers
	level	integer

```
BEGIN
    InitGauss (seed)
    FOR i := 1 TO maxlevel DO
        delta[i] := sigma * power (0.5, i*H) * sqrt (0.5) * sqrt (1 - power (2, 2*H-2))
    END FOR
    N = power (2, maxlevel)
    X[0] := 0
    X[N] := sigma * Gauss ()
    D := N
    d := D / 2
    level := 1
    WHILE (level <= maxlevel) DO
        FOR i := d TO N-d STEP D DO
            X[i] := 0.5 * (X[i-d] + X[i+d])
        END FOR
        FOR i := 0 TO N STEP d DO
            X[i] := X[i] + delta[level] * Gauss ()
        END FOR
        D := D / 2
        d := d / 2
        level := level + 1
    END WHILE
END
```

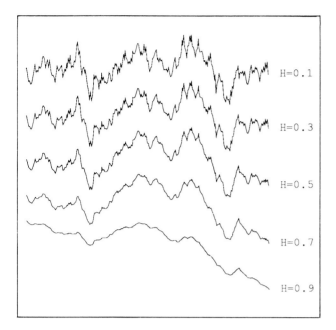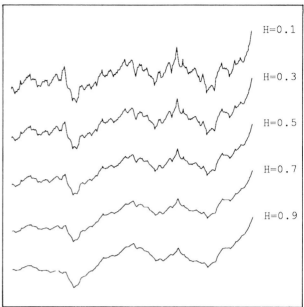

Fig. 2.7: Fractal motion generated by displacement techniques. On the left results of the algorithm *MidPointFM1D()* are shown for various parameters H. On the right the random successive additions were also included. Those curves look smoother because the amplitudes are scaled in order to fit into the graph whereas in the time direction no scaling was done. The effect is especially drastic for $H = 0.1$ where we scaled amplitudes by a factor of 0.7 and thus a properly rescaled graph would have to be contracted by a factor of $0.7^{-0.1} = 35.4$.

2.3.3 Displacing interpolated points

In the midpoint displacement method and the successive random addition method the resolution Δt is improved by a factor of $r = \frac{1}{2}$ in each stage. We can modify the second of the above methods to accommodate other factors $0 < r < 1$. For this purpose one would interpolate $X(t)$ at times $t_i = ir\Delta t$ from the samples one already has from the previous stage at a sampling rate of Δt. Then a random element D_n would be added to all of the interpolated points. In agreement with the requirements for the mean square increments of X we have to prescribe a variance proportional to $(r^n)^{2H}$ for the Gaussian random variable D_n:

$$\Delta_n^2 \propto (r^n)^{2H}. \tag{2.9}$$

The additional parameter r will change the appearance of the fractal, the feature of the fractal controlled by r has been termed the *lacunarity*. In the algorithm below we use a simple linear interpolation. The reason we include yet another method here is that it generalizes to two, three or more dimensions in a very straight forward way, while the other methods are harder to convert.

```
ALGORITHM InterpolatedFM1D  (X, N, r, sigma, H, seed)
Title            One-dimensional Fractal motion with lacunarity

Arguments    X[]          real array of size N
             N            number of elements in X
             r            scaling factor for resolutions (0 < r < 1)
             sigma        initial standard deviation
             H            0 < H < 1 determines fractal dimension D = 2 - H
             seed         seed value for random number generator
Variables    delta        real variable holding standard deviations Δ
             Y[]          real array of size N for the interpolated values of X
             mT, mt       integers, number of elements in arrays X and Y
             t, T         sampling rates for X and Y
             i, index     integer
             h            real

BEGIN                     /* initialize the array with 2 points */
    InitGauss (seed)
    X[0] := 0
    X[1] := sigma * Gauss ()
    mT := 2
    T := 1.0
                          /* loop while less than N points in array */
    WHILE (mT < N) DO
                          /* set up new resolution of mt points */
        mt := mT / r
        IF (mt = mT) THEN mt := mT + 1
        IF (mt > N) THEN mt := N
        t := 1 / (mt-1)
                          /* interpolate new points from old points */
        Y[0] := X[0]
        Y[mt-1] := X[mT-1]
        FOR i := 1 TO mt-2 DO
            index := integer (i * t / T)
            h := i * t / T - index
            Y[i] := (1 - h) * X[index] + h * X[index+1]
        END FOR
                          /* compute the standard deviation for offsets */
        delta := sqrt (0.5) * power (t, H) * sigma * sqrt (1.0 - power (t/T, 2-2*H))
                          /* do displacement at all positions */
        FOR i := 0 TO mt-1
            X[i] := Y[i] + delta * Gauss ()
        END FOR
        mT := mt
        T := 1 / mT
    END WHILE
END
```

Following the ideas for midpoint displacement for Brownian motion we set $X(0) = 0$ and select $X(1)$ as a sample of a Gaussian random variable with variance σ^2. Then we can deduce in the same fashion as before that

$$\Delta_n^2 = \frac{1}{2}(1 - r^{2-2H})(r^n)^{2H}\sigma^2.$$

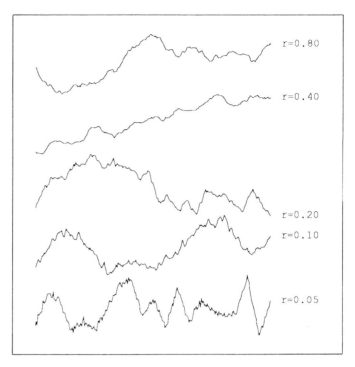

Fig. 2.8: Fractal Brownian motion with varying lacunarity. The algorithm *InterpolatedFM1D()* produced these curves with a parameter $H = 0.8$. The other parameter r is the scaling factor in the method. $r = 0.1$ means that in each stage we have $1/r = 10$ times as many points as in the previous stage. Especially for low r we see a pronounced effect of these high scaling ratios as a modulation remaining from the first one or two stages. This is called the lacunarity.

A little difficulty arises since we cannot work with a continuous variable $X(t)$ in a computer program (X is stored as an array). Thus, generally, one cannot maintain the constant factor of r reducing the resolution Δt in each step. Let us assume for simplicity that we have some $X(0)$ and $X(T)$ already computed where $T \approx r^{n-1}$. Then $X(t)$ with $t \approx r^n$ is first interpolated via

$$X(t) = X(0) + \frac{t}{T}(X(T) - X(0)).$$

Then all points $X(0), X(t), X(2t)$, etc. are displaced by samples of a Gaussian random variable D with variance Δ^2. The old values $X(0)$ and $X(T)$ satisfied

$$\text{var}(X(T) - X(0)) = T^{2H}\sigma^2,$$

and we have to select the variance Δ^2 such that

$$\text{var}(X(t) - X(0)) = t^{2H}\sigma^2$$

holds. Therefore we require

$$t^{2H}\sigma^2 = (\frac{t}{T})^2 T^{2H}\sigma^2 + 2\Delta^2$$

where the first term of the sum stands for the variance already contained in the interpolation, and the other represents the variance due to the perturbations in both $X(0)$ and $X(t)$. Thus

$$\Delta^2 = \frac{1}{2}\sigma^2(1 - (\frac{t}{T})^{2-2H})t^{2H}.$$

To apply this formula to the case of random successive additions we set $T = \frac{1}{2^{n-1}}$ and $t = \frac{1}{2^n}$, and we obtain

$$\Delta^2 = \frac{1}{2}\sigma^2(1 - \frac{1}{2^{2-2H}})\frac{1}{2^{2Hn}}$$

which is, as expected, the quantity Δ_n^2 used in the algorithm *AdditionsFM1D()* where the ratio r is $\frac{1}{2}$.

2.4   Fractional Brownian motion : Approximation by spectral synthesis

2.4.1   The spectral representation of random functions

The spectral synthesis method (also known as the Fourier filtering method) for generating fBm is based on the spectral representation of samples of the process $X(t)$. Since the Fourier transform of X generally is undefined we first restrict $X(t)$ to a finite time interval, say $0 < t < T$:

$$X(t, T) = \begin{cases} X(t) & \text{if } 0 < t < T \\ 0 & \text{otherwise} \end{cases}.$$

We thus have

$$X(t, T) = \int_{-\infty}^{\infty} F(f, T) e^{2\pi i t f} df,$$

where $F(f, T)$ is the Fourier transform of $X(t, T)$

$$F(f, T) = \int_0^T X(t) e^{-2\pi i f t} dt.$$

Now $|F(f, T)|^2 df$ is the contribution to the total energy of $X(t, T)$ from those components with frequencies between f and $f + df$. The average power of X contained in the interval $[0, T]$ is then given by

$$\frac{1}{T} \int_{-\infty}^{\infty} |F(f, T)|^2 df,$$

and the *power spectral density* of $X(t, T)$ is

$$S(f, T) = \frac{1}{T}|F(f, T)|^2.$$

The spectral density of X is then obtained in the limit as $T \to \infty$

$$S(f) = \lim_{T \to \infty} \frac{1}{T} |F(f,T)|^2 .$$

The interpretation of $S(f)$ is the following: $S(f)\,df$ is the average of the contribution to the total power from components in $X(t)$ with frequencies between f and $f + df$. $S(f)$ is a nonnegative and even function. We may think of $X(t)$ as being decomposed into a sum of infinitely many sine and cosine terms of frequencies f whose powers (and amplitudes) are determined by the spectral density $S(f)$. For a good introduction into the theory of spectral analysis of random functions we recommend [87] and (shorter) [20]. A version which is more mathematically oriented (but still accessible to readers with some knowledge of probability theory) is contained in Part I of [107].

2.4.2 The spectral exponent β in fractional Brownian motion

The underlying idea of spectral synthesis is that a prescription of the right kind of spectral density $S(f)$ will give rise to fBm with an exponent $0 < H < 1$.

If the random function $X(t)$ contains equal power for all frequencies f, this process is called white noise in analogy with the white light made up of radiations of all wave lengths. If $S(f)$ is proportional to $1/f^2$ we obtain the usual brown noise or Brownian motion. In general, a process $X(t)$ with a spectral density proportional to $1/f^\beta$ corresponds to fBm with $H = \frac{\beta-1}{2}$:

$$S(f) \propto \frac{1}{f^\beta} \sim \text{fBm with } \beta = 2H + 1. \qquad (2.10)$$

Choosing β between 1 and 3 will generate a graph of fBm with a fractal dimension of

$$D_f = 2 - H = \frac{5 - \beta}{2}. \qquad (2.11)$$

Let us pause for a moment to explain this relationship between β and H. Mathematically it can be derived from the fact that the mean square increments (which are proportional to Δt^{2H} for fBm with exponent H) are directly related to the autocorrelation function of X, which in turn defines the spectral density by means of a Fourier transform via the Wiener-Khintchine relation (see Section 1.6.8). In place of this "pure" approach we propose a simple and more heuristic argument for the relation $\beta = 2H + 1$. We start out by restating the fundamental property of fBm: If $X(t)$ denotes fBm with exponent $0 < H < 1$ then the properly rescaled random function

$$Y(t) = \frac{1}{r^H} X(rt)$$

for a given $r > 0$ has the same statistical properties as X. Thus it also has the same spectral density. From this basic observation we can deduce the important result (2.10) using only the above definitions and some elementary calculus as follows.

Let us fix some $r > 0$, set

$$Y(t,T) = \begin{cases} Y(t) = \frac{1}{r^H}X(rt), & \text{if } 0 < t < T \\ 0, & \text{otherwise} \end{cases}$$

and adopt the notation

$$
\begin{array}{ll}
F_X(t,T), F_Y(t,T) & \text{Fourier transforms of } X(t,T), Y(t,T), \\
S_X(f,T), S_Y(f,T) & \text{spectral densities of } X(t,T), Y(t,T), \\
S_X(f), S_Y(f) & \text{spectral densities of } X(t), Y(t).
\end{array}
$$

We compute

$$F_Y(f,T) = \int_0^T Y(t)e^{-2\pi ift}dt = \frac{1}{r^H}\int_0^{rT} X(s)e^{-2\pi i\frac{f}{r}s}\frac{ds}{r},$$

where we have substituted $\frac{s}{r}$ for t and $\frac{ds}{r}$ for dt in the second integral. Thus, clearly

$$F_Y(f,T) = \frac{1}{r^{H+1}}F_X(\frac{f}{r}, rT).$$

Now it follows for the spectral density of $Y(t,T)$

$$S_Y(f,T) = \frac{1}{r^{2H+1}}\frac{1}{rT}|F_X(\frac{f}{r}, rT)|^2$$

and in the limit as $T \to \infty$ or equivalently as $rT \to \infty$ we conclude

$$S_Y(f) = \frac{1}{r^{2H+1}}S_X(\frac{f}{r}).$$

Since Y is just a properly rescaled version of X, their spectral densities must coincide, thus, also

$$S_X(f) = \frac{1}{r^{2H+1}}S_X(\frac{f}{r}).$$

Now we formally set $f = 1$ and replace $1/r$ again by f to finally obtain the desired result (2.10)

$$S_X(f) \propto \frac{1}{f^{2H+1}} = \frac{1}{f^\beta}.$$

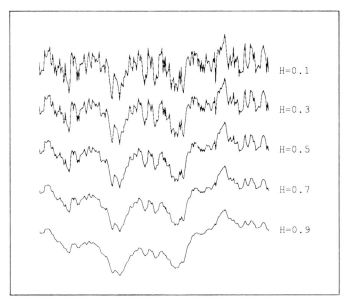

Fig. 2.9: Fractal motion via spectral synthesis. The above curves correspond to spectral density functions of the form $1/f^\beta$ where $\beta = 2H + 1$. The fractal dimensions of these graphs are $2 - H$.

2.4.3 The Fourier filtering method

For the practical algorithm we have to translate the above into conditions on the coefficients a_k of the discrete Fourier transform

$$\overline{X}(t) = \sum_{k=0}^{N-1} a_k e^{2\pi ikt} \qquad (2.12)$$

The coefficients a_k are in a 1:1 correspondence with the complex values $\overline{X}(t_k)$, $t_k = \frac{k}{N}, k = 0, 1, \ldots, N-1$. The condition to be imposed on the coefficients in order to obtain $S(f) \propto \frac{1}{f^\beta}$ now becomes

$$E(|a_k|^2) \propto \frac{1}{k^\beta} \qquad (2.13)$$

since k essentially denotes the frequency in (2.12). This relation (2.13) holds for $0 < k < \frac{N}{2}$. For $k \geq \frac{N}{2}$ we must have $a_k = \overline{a_{N-k}}$ because $\overline{X}$ is a real function. The method thus simply consists of randomly choosing coefficients subject to the expectation in (2.13) and then computing the inverse Fourier transform to obtain X in the time domain. Since the process X need only have real values it is actually sufficient to sample real random variables A_k and B_k under the constraint

$$E(A_k^2 + B_k^2) \propto \frac{1}{k^\beta}$$

and then set

$$\overline{X}(t) = \sum_{k=1}^{N/2}(A_k \cos kt + B_k \sin kt).\qquad(2.14)$$

In contrast to some of the previous algorithms discussed this method is not recursive and also does not proceed in stages of increasing spatial resolution. We may, however, interpret the addition of more and more random Fourier coefficients a_k satisfying (2.13) as a process of adding higher frequencies, thus, increasing the resolution in the frequency domain.

```
ALGORITHM SpectralSynthesisFM1D  (X, N, H, seed)
Title           Fractal motion using Fourier filtering method

Arguments   X[]       array of reals of size N
            N         size of array X
            H         0 < H < 1 determines fractal dimension D = 2 − H
            seed      seed value for random number generator
Globals     Arand     rand() returns values between 0 and Arand
Variables   i         integer
            beta      exponent in the spectral density function (1 < beta < 3)
            rad, phase  polar coordinates of Fourier coefficient
            A[], B[]  real and imaginary parts of Fourier coefficients
Subroutines InvFFT    fast inverse Fourier transform in 1 dimension

BEGIN
    InitGauss (seed)
    beta := 2 * H + 1
    FOR i := 0 TO N/2-1 DO
        rad := power (i+1, -beta/2) * Gauss ()
        phase := 2 * 3.141592 * rand () / Arand
        A[i] := rad * cos (phase)
        B[i] := rad * sin (phase)
    END FOR
    InvFFT (A, B, X, N/2)
END
```

The advantage of this straight forward method is that it is the purest interpretation of the concept of fractional Brownian motion. Artifacts such as those occurring in the midpoint displacement methods are not apparent. However, due to the nature of Fourier transforms, the generated samples are periodic. This is sometimes annoying, and in this case one can compute twice or four times as many points as actually needed and then discard a part of the sequence.

The algorithm for the inverse Fourier transformation *InvFFT* is not included with the pseudo code *SpectralSynthesisFM1D()*. It should compute the sums (2.14) from the given coefficients in A and B. Usually fast Fourier transforms are employed. They require $O(N \log N)$ operations per transform.

2.5 Extensions to higher dimensions

2.5.1 Definitions

In this section we discuss how one can generalize the displacement methods and the spectral synthesis methods to two and three dimensions.

The generalization of fractional Brownian motion itself is straight forward. It is a multidimensional process (a random field) $X(t_1, t_2, \ldots, t_n)$ with the properties:

(i) The increments $X(t_1, t_2, \ldots, t_n) - X(s_1, s_2, \ldots, s_n)$ are Gaussian with mean 0

(ii) The variance of the increments $X(t_1, t_2, \ldots, t_n) - X(s_1, s_2, \ldots, s_n)$ depends only on the distance

$$\sqrt{\sum_{i=1}^{n} (t_i - s_i)^2}$$

and in fact is proportional to the 2H-th power of the distance, where the parameter H again satisfies $0 < H < 1$. Thus,

$$E(|X(t_1, t_2, \ldots, t_n) - X(s_1, s_2, \ldots, s_n)|^2) \propto (\sum_{i=1}^{n} (t_i - s_i)^2)^H. \tag{2.15}$$

The random field X again has stationary increments and is isotropic, i.e. all points $(t_1, t_2, \ldots, t_n)$ and all directions are statistically equivalent. In the frequency domain we have for the spectral density

$$S(f_1, \ldots, f_n) \propto \frac{1}{\left(\sqrt{\sum_{i=1}^{n} f_i^2}\right)^{2H+n}}. \tag{2.16}$$

This fact can be deduced in the exact same fashion as in the last section for the $\frac{1}{f^\beta}$ law in the one-dimensional case. Therefore we skip these details here. This ensures that X restricted to any straight line will be a $\frac{1}{f^\beta}$ noise corresponding to $2H = \beta - 1$ [103]. In analogy with formula (2.10) the fractal dimension of the graph of a sample of $X(t_1, t_2, \ldots, t_n)$ is

$$D = n + 1 - H. \tag{2.17}$$

2.5.2   Displacement methods

The midpoint displacement methods can work with square lattices of points. If the mesh size δ denotes the resolution of such a grid, we obtain another square grid of resolution $\frac{\delta}{\sqrt{2}}$ by adding the midpoints of all squares. Of course, the orientation of the new square lattice is rotated by 45 degrees. Again adding the midpoints of all squares gives us the next lattice with the same orientation as the first one and the resolution is now $\frac{\delta}{2}$ (see Figure 2.10). In each stage we thus scale the resolution with a factor of $r = \frac{1}{\sqrt{2}}$, and in accordance with (2.15) we add random displacements using a variance which is r^{2H} times the variance of the previous stage. If we assume that the data on the four corner points of the grid carry mean square increments of σ^2 then at stage n of the process we must add a Gaussian random variable of variance $\sigma^2 r^{2Hn} = \sigma^2 (\frac{1}{2})^{nH}$. For the usual midpoint method random elements are added only to the new midpoints in each stage, whereas in the random addition method we add displacements at all points. Thus we can unify both methods in just one algorithm.

In Figures 2.11 to 2.13 we show topographical maps of random fractal landscapes which were generated using the algorithm *MidPointFM2D()*. Random additions and differing seed values were used, and the parameter H varies from 0.8 to 0.2. At a resolution of 65 by 65 points (*maxlevel* = 6) a total of 4225 data points were generated. The heights of the surface were scaled to the range from -10000 to $+10000$ and then handed to a contour line program which produced the maps. The parts of the surfaces that have a negative height are assumed to be submerged under water and are not displayed. The fractal dimension of these surfaces is $3 - H$. A rendering of a perspective view of these landscapes is also included.

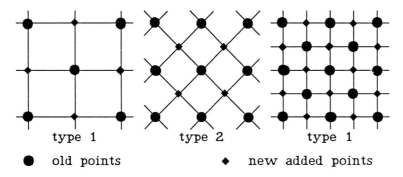

Fig. 2.10: Grid types for displacement methods. In the displacement methods a grid of type 1 is given at the beginning from which a type 2 grid is generated. Its mesh size is $1/\sqrt{2}$ times the old mesh size. In a similar step a grid of type 1 is again obtained as shown in the figure.

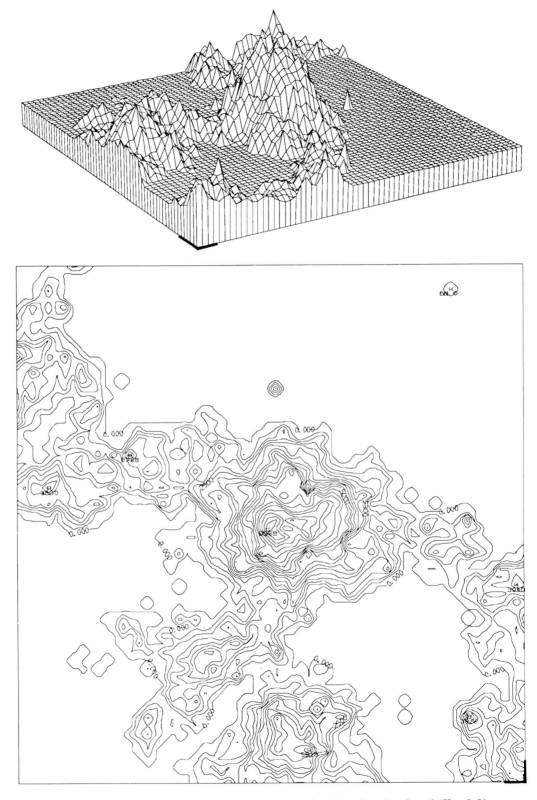

Fig. 2.11: Topographical map and perspective view of random fractal ($H = 0.8$).

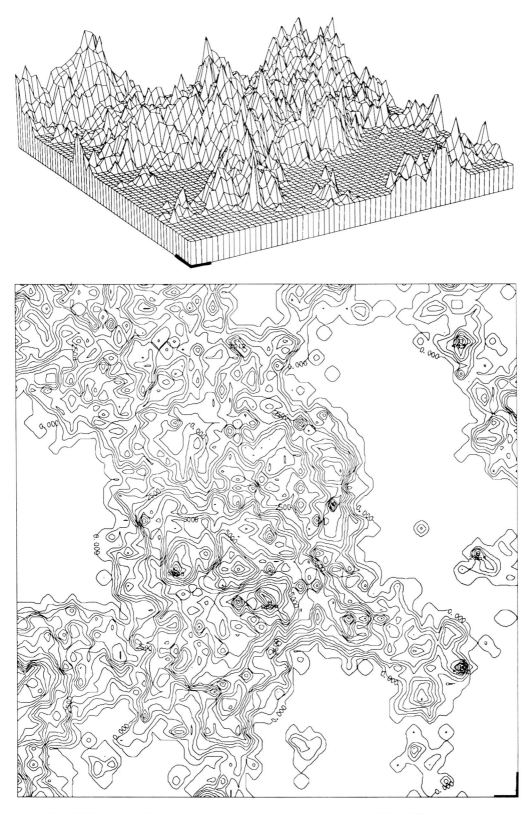

Fig. 2.12: Topographical map and perspective view of random fractal ($H = 0.5$).

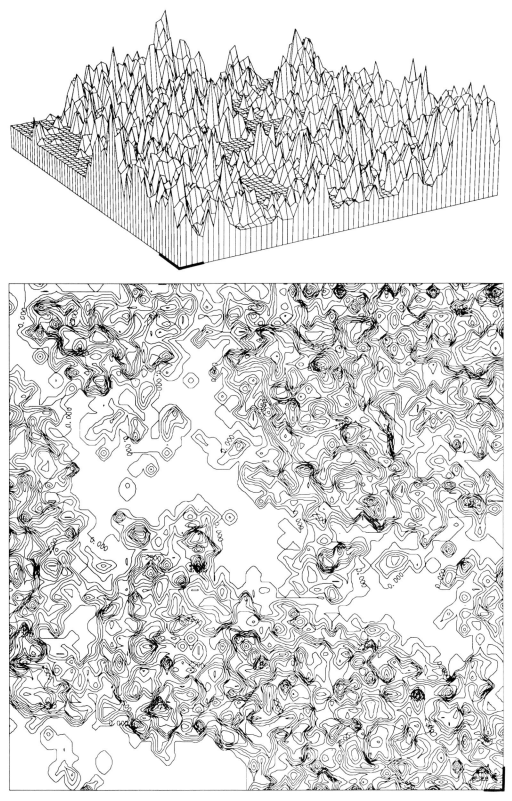

Fig. 2.13: Topographical map and perspective view of random fractal ($H = 0.2$).

ALGORITHM **MidPointFM2D** (X, maxlevel, sigma, H, addition, seed)
Title   Midpoint displacement and successive random additions in 2 dimensions

---

Arguments   X[][]   doubly indexed real array of size $(N + 1)^2$
             maxlevel   maximal number of recursions, $N = 2^{maxlevel}$
             sigma   initial standard deviation
             H   parameter H determines fractal dimension $D = 3 - H$
             addition   boolean parameter (turns random additions on/off)
             seed   seed value for random number generator
Variables   i, N, stage   integers
             delta   real holding standard deviation
             x, y, y0, D, d integer array indexing variables
Functions   f3(delta,x0,x1,x2) = (x0+x1+x2)/3 + delta * Gauss()
             f4(delta,x0,x1,x2,x3) = (x0+x1+x2+x3)/4 + delta * Gauss()

```
BEGIN
    InitGauss (seed)
    N = power (2, maxlevel)
                              /* set the initial random corners */
    delta := sigma
    X[0][0] := delta * Gauss ()
    X[0][N] := delta * Gauss ()
    X[N][0] := delta * Gauss ()
    X[N][N] := delta * Gauss ()
    D := N
    d := N / 2
    FOR stage := 1 TO maxlevel DO
                              /* going from grid type I to type II */
        delta := delta * power (0.5, 0.5*H)
                              /* interpolate and offset points */
        FOR x :=d TO N-d STEP D DO
            FOR y := d TO N-d STEP D DO
                X[x][y] := f4 (delta, X[x+d][y+d], X[x+d][y-d], X[x-d][y+d],
                    X[x-d][y-d])
            END FOR
        END FOR
                              /* displace other points also if needed */
        IF (addition) THEN
            FOR x := 0 TO N STEP D DO
                FOR y := 0 TO N STEP D DO
                    X[x][y] := X[x][y] + delta * Gauss ()
                END FOR
            END FOR
        END IF
        (to be continued on the next page)
```

We also remark that one does not necessarily have to work with rectangular grids. Some authors also consider triangular subdivisions (see [40,49,50,76] and Appendix A). For a detailed discussion of displacement techniques, their history and possible extensions, we refer to Appendix A.

The algorithm *InterpolatedFM1D()* of Section 2.3 implementing variable scaling factors of resolutions can be modified to compute approximations of

<pre>
ALGORITHM MidPointFM2D (X, maxlevel, sigma, H, addition, seed)
Title   Midpoint displacement and successive random additions in 2 dimensions
               (continued from previous page)
</pre>

```
                          /* going from grid type II to type I */
          delta := delta * power (0.5, 0.5*H)
                             /* interpolate and offset boundary grid points */
          FOR x := d TO N-d STEP D DO
              X[x][0] := f3 (delta, X[x+d][0], X[x-d][0], X[x][d])
              X[x][N] := f3 (delta, X[x+d][N], X[x-d][N], X[x][N-d])
              X[0][x] := f3 (delta, X[0][x+d], X[0][x-d], X[d][x])
              X[N][x] := f3 (delta, X[N][x+d], X[N][x-d], X[N-d][x])
          END FOR
                             /* interpolate and offset interior grid points */
          FOR x := d TO N-d STEP D DO
              FOR y := D TO N-d STEP D DO
                  X[x][y] := f4 (delta, X[x][y+d], X[x][y-d], X[x+d][y], X[x-d][y])
              END FOR
          END FOR
          FOR x := D TO N-d STEP D DO
              FOR y := d TO N-d STEP D DO
                  X[x][y] := f4 (delta, X[x][y+d], X[x][y-d], X[x+d][y], X[x-d][y])
              END FOR
          END FOR
                             /* displace other points also if needed */
          IF (addition) THEN
              FOR x := 0 TO N STEP D DO
                  FOR y := 0 TO N STEP D DO
                      X[x][y] := X[x][y] + delta * Gauss ()
                  END FOR
              END FOR
              FOR x := d TO N-d STEP D DO
                  FOR y := d TO N-d STEP D DO
                      X[x][y] := X[x][y] + delta * Gauss ()
                  END FOR
              END FOR
          END IF
          D := D / 2
          d := d / 2
      END FOR
END
```

fBm in two or three (or even higher) dimensions. If d denotes this dimension we have to fill an array of size N^d elements subject to (2.15). These points are evenly spaced grid points in a unit cube, thus the final resolution (grid size) is $1/(N-1)$. At a particular stage of this algorithm we have an approximation at resolution $1/(M-1)$ given in the form of an array of size M^d with $M < N$. The next stage is approached in two steps: 1. If $0 < r < 1$ denotes the factor at which the resolution changes, then a new approximation consisting of L^d points where $L \approx M/r$ must be computed. First the values of these numbers are taken as multilinear or higher order interpolation of the M^d old values.

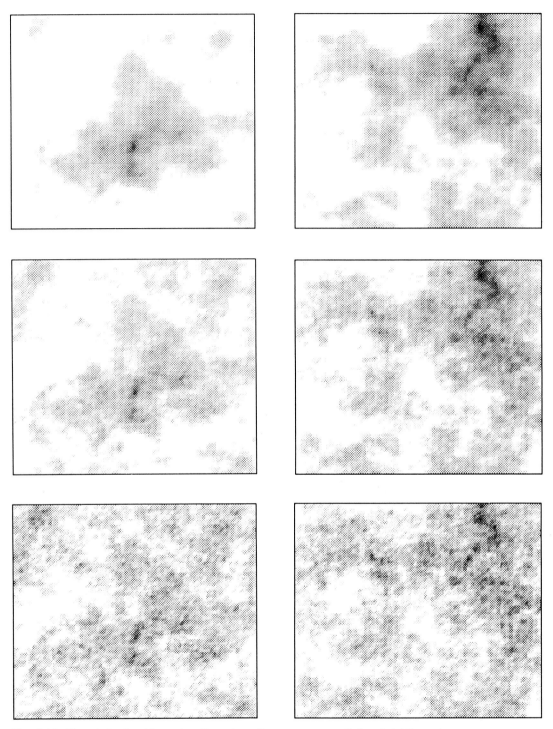

Fig. 2.14: Fractal clouds with varying dimensions. In two sequences (left and right) we show the effect of the parameter H. The data for the clouds was produced in the same way as in the Figures 2.11 to 2.13 (the parameter maxlevel was at 7 though). The rendering displays a top view of the "landscape" where the height now defines the shade of gray to be used. Points with heights below zero are not drawn, i.e. they appear as white (see Section 2.7.1). The parameter H varies from 0.8 (top) to 0.5 (middle) to 0.2 (bottom).

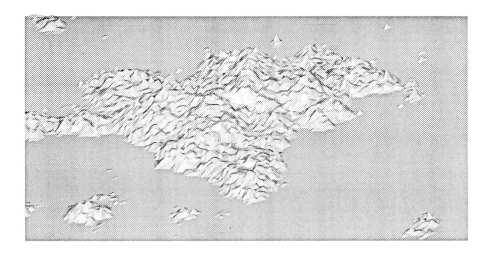

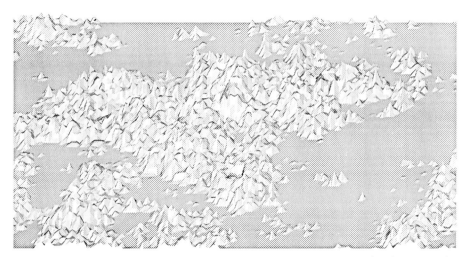

Fig. 2.15: Clouds as mountains. The two first clouds of Figure 2.14 are rendered as mountains here.

2. A random element of variance proportional to $1/(L-1)^{2H}$ is added to all points. These two steps are repeated until we have obtained N^d points.

We briefly describe the multilinear interpolation (compare [86]). Let us first take $d = 2$ and assume that a grid point (x, y) of the new grid falls into a square of old grid points $(x_0, y_0), (x_0, y_1), (x_1, y_0), (x_1, y_1)$ (see Figure 2.16). We can express (x, y) relative to these points by

$$
\begin{aligned}
x &= (1 - h_x)x_0 + h_x x_1 \\
y &= (1 - h_y)y_0 + h_y y_1
\end{aligned}
$$

where $0 \le h_x, h_y \le 1$. If $X(x_i, y_j)$ denotes the approximation of fBm at the

---

ALGORITHM **InterpolatedFM** (X, N, d, r, sigma, H, seed)
Title   Fractal motion via interpolation and variable scaling

---

Arguments	X[]	real array of size N^d
	N	number of elements in X along one dimension
	d	dimension of the space (1, 2, 3 or more)
	r	scaling factor for resolutions $(0 < r < 1)$
	sigma	initial standard deviation
	H	parameter $0 < H < 1$ determines fractal dimension
	seed	seed value for random number generator
Variables	delta	real variable holding standard deviations Δ
	Y[]	real array of size N for the interpolated values of X
	mT, mt	integers, number of elements N^d in arrays X and Y along one dimension
	t, T	real sampling rates for X and Y
	i, index	integer
	h	real
Subroutine	Interpolate()	multilinear interpolation routine (see text)

```
BEGIN
                        /* initialize the array with 2^d points */
    InitGauss (seed)
    mT := 2
    FOR i := 0 TO power(2,d)-1 DO
        X[i] := sigma * Gauss ()
    END FOR
                        /* loop while less than N^d points in array */
    WHILE (mT < N) DO
                        /* set up new resolution of mt points */
        mt := mT / r
        IF (mt = mT) THEN mt := mT + 1
        IF (mt > N) THEN mt := N
        t := 1.0 / (mt-1)
        T := 1.0 / (mT-1)
                        /* interpolate new points from old points */
        Interpolate (X, mT, Y, mt, d)
                        /* compute the standard deviation for offsets */
        delta := sqrt (0.5) * sigma * sqrt (1.0 - power (t/T, 2-2*H)) * power (t, H)
                        /* do displacement at all positions */
        FOR i := 0 TO power(mt,d)-1 DO
            X[i] := Y[i] + delta * Gauss ()
        END FOR
        mT := mt
    END WHILE
END
```

---

old points we now interpolate at the point (x, y) by setting

$$
\begin{aligned}
X(x, y) \;=\; & (1 - h_x)(1 - h_y) X(x_0, y_0) + h_x(1 - h_y) X(x_1, y_0) \\
+\; & (1 - h_x) h_y X(x_0, y_1) + h_x h_y X(x_1, y_1).
\end{aligned}
$$

In three dimensions $(d = 3)$ we have (x, y, z) with x, y as above and $z = (1 - h_z) z_0 + h_z z_1$ and the above interpolation formula extends to this case in

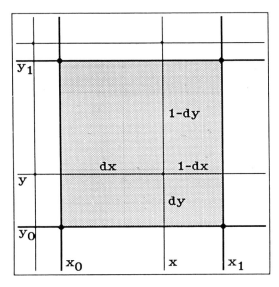

Fig. 2.16: Multilinear interpolation (see text).

the obvious way. In the pseudo code *InterpolatedFM()* we do not include these very technical details. The points are stored in a long linear array X of floating point numbers and the routine *Interpolate* should be the implementation of the multilinear interpolation. Its inputs are the array X, the number of points along one of the dimensions, the dimension, and the requested number of points in one dimensions. The output is a new array Y with the specified size.

2.5.3 The Fourier filtering method

We now proceed to the extension of the Fourier filtering method to higher dimensions. In two dimensions the spectral density S generally will depend on two frequency variables u and v corresponding to the x and y directions. But since all directions in the xy-plane are equivalent with respect to statistical properties, S depends only on $\sqrt{u^2 + v^2}$. If we cut the surface along a straight line in the xy-plane, we expect for the spectral density S of this fBm in only one dimension a power law $1/f^\beta$ as before. This requirement implies (see (2.16)) for the two-dimensional spectral density to behave like

$$S(u,v) = \frac{1}{(u^2 + v^2)^{H+1}}.$$

The two-dimensional discrete inverse Fourier transform is

$$X(x,y) = \sum_{k=0}^{N-1}\sum_{l=0}^{N-1} a_{kl}e^{2\pi i(kx+ly)}$$

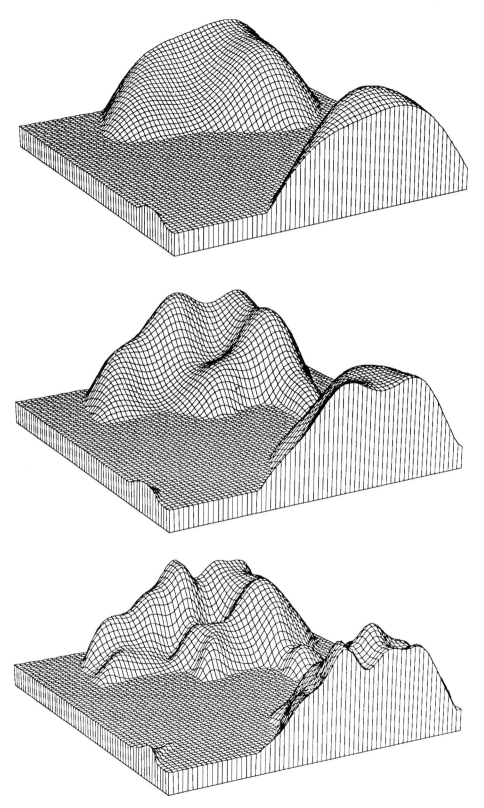

Fig. 2.17:   Spectral synthesis of a mountain.

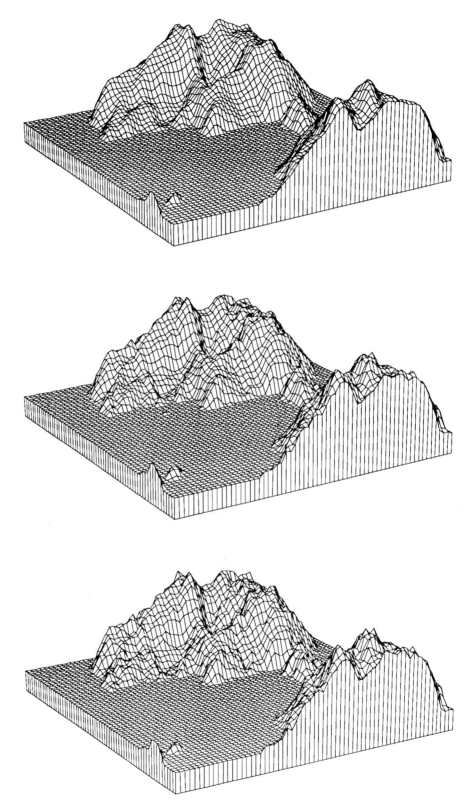

Fig. 2.18: Spectral synthesis of a mountain (continued).

ALGORITHM **SpectralSynthesisFM2D** (X, N, H, seed)
Title   Fractal motion in 2 dimensions

Arguments	X[][]	doubly indexed array of complex variables of size N^2
	N	size of array X along one dimension
	H	$0 < H < 1$ determines fractal dimension D = 3 - H
	seed	seed value for random number generator
Globals	Arand	rand() returns values between 0 and Arand
Variables	i, j, i0, j0	integers
	rad, phase	polar coordinates of Fourier coefficient
	A[][]	doubly indexed array of complex variables of size N^2
Subroutines	InvFFT2D	fast inverse Fourier transform in 2 dimensions
	rand()	returns system random numbers

```
BEGIN
    InitGauss (seed)
    FOR i = 0 TO N/2 DO
        FOR j = 0 TO N/2 DO
            phase := 2 * 3.141592 * rand () / Arand
            IF (i≠0 OR j≠0) THEN
                rad := power(i*i+j*j,-(H+1)/2) * Gauss()
            ELSE
                rad := 0
            END IF
            A[i][j] := (rad * cos(phase), rad * sin(phase))
            IF (i=0) THEN
                i0 := 0
            ELSE
                i0 := N - i
            END IF
            IF (j=0) THEN
                j0 := 0
            ELSE
                j0 := N - j
            END IF
            A[i0][j0] := (rad * cos(phase), - rad * sin(phase))
        END FOR
    END FOR
    A[N/2][0].imag := 0
    A[0][N/2].imag := 0
    A[N/2][N/2].imag := 0
    FOR i = 1 TO N/2-1 DO
        FOR j = 1 TO N/2-1 DO
            phase := 2 * 3.141592 * rand() / Arand
            rad := power(i*i+j*j,-(H+1)/2) * Gauss()
            A[i][N-j] := (rad * cos(phase), rad * sin(phase))
            A[N-i][j] := (rad * cos(phase), - rad * sin(phase))
        END FOR
    END FOR
    InvFFT2D(A,X,N)
END
```

for $x, y = 0, \frac{1}{N}, \frac{2}{N}, \ldots, \frac{N-1}{N}$ (see [46]), and thus, we specify for the coefficients a_{kl}

$$E(|a_{kl}|^2) \propto \frac{1}{(k^2 + l^2)^{H+1}}.$$

Since the constructed function X is a real function we must also satisfy a conjugate symmetry condition, namely

$$
\begin{aligned}
a_{N-i,N-j} &= \overline{a_{i,j}} && \text{for} \quad i,j > 0, \\
a_{0,N-j} &= \overline{a_{0,j}} && \text{for} \quad j > 0, \\
a_{N-i,0} &= \overline{a_{i,0}} && \text{for} \quad i > 0, \\
a_{0,0} &= \overline{a_{0,0}}.
\end{aligned}
$$

The fractal dimension D_f of the surface will be $D_f = 3 - H$. The algorithm then consists simply of choosing the Fourier coefficients accordingly and then performing a two-dimensional inverse transform.

The code for the inverse Fourier transformation is not included. It can easily be constructed from the one-dimensional version *InvFFT* (see [46]). As in the one-dimensional case one can reduce the number of coefficients since the random function is only real. The method also generalizes to three dimensions (see Chapter 1).

In the sequence of Figure 2.17 and 2.18 we show how the spectral synthesis method "adds detail" by improving the spectral representation of the random fractal, i.e. by allowing more and more Fourier coefficients to be employed. The resolution in the algorithm *SpectralSynthesisFM2D()* was $N = 64$ but in the top image of Figure 2.17 only the first $2^2 = 4$ coefficients were used. In the other pictures we allowed $4^2 = 16$ (middle) and $8^2 = 64$ (bottom) non-zero coefficients. In Figure 2.18 we increase the number of Fourier coefficients to 16^2 (top), 32^2 (middle) and 64^2 (bottom).

2.6 Generalized stochastic subdivision and spectral synthesis of ocean waves

A glance at the SIGGRAPH proceedings of the last few years will reveal that the computer graphics community is very interested in the rendering aspect of fractals and other procedurally defined objects. In particular ray tracing methods have been adapted to handle the infinite detail at small scales that is characteristic for fractals. The images of random fractals presented by Voss demonstrate the perfection with which it is now possible to generate mountain scenes, craters on a moon etc. The question remains in what way the methods used for the production of random fractals can be modified in order to model even more natural phenomena. In this last section we will briefly review two such approaches published in the recent papers [60] and [74]. Both papers pursue the idea of spectral synthesis.

J.P. Lewis generalizes the displacement methods in his paper [60] to provide a better control of the synthesized random functions. He calls it *generalized stochastic subdivision*. His generated noises are stationary with a prescribed autocorrelation function

$$R(s) = E(X(t)X(t+s)).$$

This information is equivalent to the spectral density function since the two functions are a Fourier transform pair as stated by the Wiener-Khintchine relation. In the usual displacement techniques a new point is first obtained via interpolation from its immediate neighbors and then displaced by a random element. The generalization also has to take points into account that are not immediate neighbors. In the one-dimensional case, if X is known at times $t + t_k$, then sums of the form

$$\hat{X}(t) = \sum_k a_k X(t + t_k)$$

yield a linear estimator for the random function X at time t. In the usual midpoint technique, valid for spectral density functions $S(f) \propto \frac{1}{f^\beta}$, this sum just ranges over two points, and the weights are both equal to $\frac{1}{2}$. In the generalized stochastic subdivision the weights a_k have to be chosen in harmony with the prescribed autocorrelation function $R(s)$ or equivalently with the spectral density function $S(f)$ because in the general case the expected value of X at a midpoint no longer is just the average of its two neighbors. The values of the coefficients are deduced using an orthogonality principle known in estimation theory (see the references in [60]). Eventually this amounts to solving a linear system of equations at each stage of the subdivision method. Of course, the estimated value $\hat{X}(t)$ is finally displaced by a random element with a properly chosen variance. Following Lewis, some of the expected advantages of this method over the direct synthesis via the spectral density function and inverse Fourier transformation are (1) that the resolution of synthesized functions may be adjusted to the resolution required by its local perspective projection in the rendering process, (2) that portions of the noise may be reproduced at different locations and at different resolutions, and (3) that the random function can be required to pass through certain points, e.g. to ensure boundary continuity with adjacent objects. Some of the considered spectra are the Markovian-like spectrum $1/(a^2 + f^2)^d$, where the parameter d is similar to H in the random fractals (set $2d = 2H + 1$), and the Lorentzian spectrum, $1/(a^2 + b^2(f - f_0)^2)$. In two dimensions spectra with directional trends and others yielding oscillatory textures are easily accommodated in the setup of generalized stochastic subdivision.

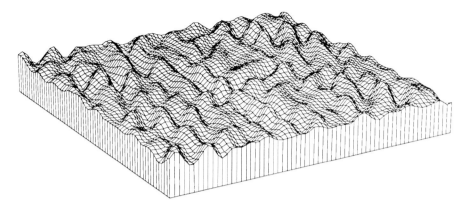

Fig. 2.19: Nonfractal waves. The spectral synthesis method was applied using a spectral density function similar to (2.18). These waves are not fractals since the spectral density falls off as $1/f^5$ for high frequencies.

The objective of the other paper by G. Mastin et al. [74] is to model a "fully developed sea in nature" using the spectral synthesis method with a spectral density function based on empirical results. Earlier researchers had recorded measurements of waves leading to a spectral model first in the downwind direction and then also in other directions. The resulting two-dimensional spectrum of these models, called the modified Pierson-Moskowics spectrum, can be formulated as

$$S(f, \phi) = \frac{a}{f^5} e^{-(\frac{b}{f})^4} D(f, \phi).$$ (2.18)

Here (f, ϕ) are the polar coordinates of the frequency variables; a and b are certain positive constants;

$$D(f, \phi) = \frac{1}{N(f)} (\cos(\frac{\phi}{2}))^{2p(f)}$$

is a directional spreading factor that weighs the spectrum at angles ϕ from the downwind direction; $p(f)$ is a piecewise power function of the type

$$p(f) = \begin{cases} a_1 (f/f_m)^{b_1} & \text{if } f < f_m \\ a_2 (f/f_m)^{-b_2} & \text{if } f \geq f_m \end{cases}$$

where f_m, a_1, a_2, b_1 and b_2 are positive constants; and $N(f)$ is chosen such that

$$\int_{-\pi}^{\pi} D(f, \phi) \, d\phi = 1.$$

There are numerous parameters and constants that enter into this formulation, e.g. the peak frequency f_m, the gravitational constant g and the wind speed at a given height above sea level.

For the implementation the given spectral density determines the expected magnitude of the complex Fourier coefficients of the sea model. In the paper this is achieved by filtering a white noise image appropriately, i.e. by multiplying the squared magnitudes of the Fourier transform of the white noise with the spectral density function. A following inverse transform then yields the sea model as a height function over the xy-plane. Plate 16 shows the result in a raytraced image with an added maehlstrom.

2.7   Computer graphics for smooth and fractal surfaces

The output of the algorithms for the synthesis of fractal surfaces (*MidPoint-FM2D()*, *InterpolatedFM()* and *SpectralSynthesisFM2D()*) is given in the form of an array which holds surface elevation data for a rectangular or square region. This data can be graphically represented in various ways. We can distinguish two cases depending on the relation between the screen resolution and the computed resolution of the surface (number of points). If the number of computed elevations is large enough, i.e. approximately equal to the number of pixels on the screen, we can use these points as primitives for the rendering. In the other case we have to interpolate the surface between data points e.g. by using polygons or bilinear patches. Ray tracings have added another layer of realism to fractals. We forgo these methods and refer the reader to [14,18,58,76].

2.7.1   Top view with color mapped elevations

This is the simplest way to render a picture of the data. Geometrically the surface is treated as flat, and the elevation of the surface is color coded so that e.g. the low regions correspond to various shades of blue (depending on depth) and the high elevations are shown in shades of green, brown and white. In this way one can produce the equivalent of a geographical map of the landscape. If there are enough elevation data points we draw one pixel of color per point. Otherwise we can use the interpolation/multilinear of Section 2.5 to obtain as many new points as needed. E.g. for a screen resolution of 768 by 1024 an array of 192 by 256 elevations is sufficient. By multilinear interpolation the resolution can be quadrupled in the horizontal and vertical direction thus matching the given screen resolution. This procedure may be considered a special case of Gouraud shading. Color Plate 14 has been produced this way, the data were computed with *MidPointFM2D()*.

By changing the colors corresponding to various heights it is very easy to produce convincing clouds. Color Plate 15 shows the same data as Plate 14 but uses a color map as follows: We specify colors in RGB space (red, green, blue) with intensities varying from 0 to 255. Let h_{min} and h_{max} denote the minimum and maximum heights of the data and $\overline{h} = \frac{1}{2}(h_{min} + h_{max})$. Then the color corresponding to height h with $h_{min} < h < h_{max}$ is defined as follows : If $h \leq \overline{h}$, then the color values for red and green are 0, where the blue value is 255. If $h > \overline{h}$, then the value of the blue component remains at 255, while the red and green portions are given by

$$255 \cdot \frac{h - \overline{h}}{h_{max} - \overline{h}}.$$

This defines a color map which is blue in the lower half and is a color ramp from blue to white in the upper half. This method to generate pictures of clouds of course does not model the natural phenomenon "clouds" physically accurately. For example, the three-dimensionality is not taken into account.

In the above procedure the geometry of the scene is rather trivial, the family of displayed polygons forms a flat square or rectangle. The elevation data is transformed into color. The obvious extension of this is to use the elevation data as the third coordinate for the vertices of the polygons. This turns the flat surface into fractal terrain embedded in three-dimensional space. One then has to define a projection of the polygons onto the screen. The painter's algorithm sorts the polygons according to decreasing distance, and then draws them from back to front, thereby eliminating the hidden parts of the surface. This and the application of a lighting model is a standard procedure (see e.g. [39]). Figure 2.15 is a coarse example. In addition to the three-dimensionality of the surface we again can use the elevation data to define colors relative to height. The lighting model then adds intensities.

2.7.2   Extended floating horizon method

All methods for the approximation of two-dimensional fractional Brownian motion easily generate as many points as wished. Using the Fourier synthesis method, however, it may take a longer time to perform the compute intensive 2D Fourier transformations. If enough points are produced it is possible in the rendering to avoid the interpolation that is introduced by using polygons. The points themselves serve as primitives for the rendering. This approach obviously takes the idea of fractals more seriously, since the characteristic of fractals, that

Color plates and captions

In the following we list the captions for the color plates of the following pages. Captions for the front and back cover images are at the end of this list. At the end of this section we include credits for the authors of the images.

Plates 1 to 7 show the stages of creating the fractal planet by adding random faults encircling the sphere.

Plate 1 The surface height variations (as indicated by color) after 10 faults shown both on the sphere and on a flat projection map.

Plate 2 The surface map after 20 faults where the colors indicate different land altitudes and ocean depths.

Plate 3 The surface map after 100 faults.

Plate 4 The surface map after 500 faults. Here the land has condensed into a single mass similar to the earth before continental drift.

Plate 5 The surface after more than 10000 faults with added polar caps. Additional faults would cause changes of detail but the overall appearance would remain the same.

Plate 6 This Brownian fractal planet mapped back onto the sphere (surface $D = 2.5$) with an earth-like water level and polar caps.

Plate 7 The same planet without water.

The differing visual effect of changing the fractal dimension D for the same landscape is shown in Plates 11 to 13. As D increases, the perceived surface roughness also increases. Also shown is fractal landscaping by scaling the surface height variations (relative to water level) by a power law in Plates 8 to 10.

Plate 8 $D = 2.15$, the cube root of the original height is taken.

Plate 9 $D = 2.15$, original height.

Plate 10 $D = 2.15$, cubed height.

Plate 11 $D = 2.15$

Plate 12 $D = 2.5$

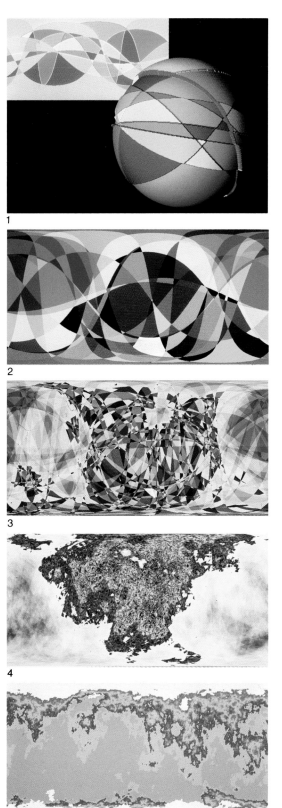

1

2

3

4

5

6

7

8

9

10

11

12

13

14

15

16a

16b

16c

16d

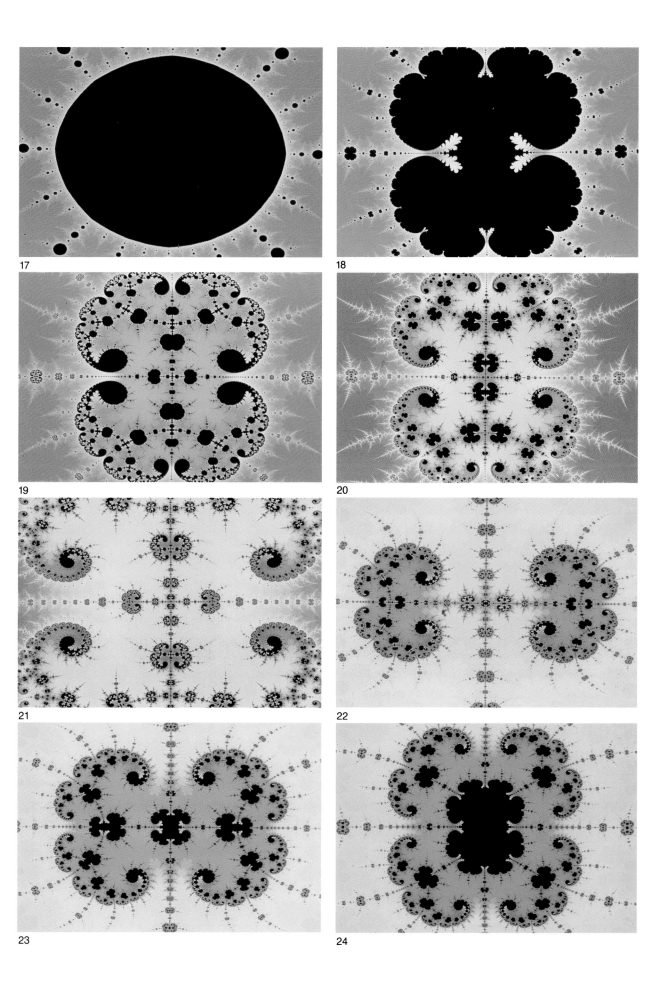

17

18

19

20

21

22

23

24

25

26

27

28

29

30

31

32

33

34

35

36

Plate 13 $D = 2.8$

Plate 14 Top view of fractal landscape with color coded heights. Blue shades indicate elevation below 0 and green, brown, white corresponds to increasing heights above 0. Data was generated using *MidPointFM2D()*.

Plate 15 Same fractal data as in Plate 14 with different color map applied to surface elevation yielding simulated clouds.

Plate 16 The Fourier synthesis method is applied to generate ocean waves, mountains and the sky in image (a), called *Drain*. Stochastic subdivision is used in the sunset picture (b), see Section 2.6 for methods used in (a) and (b). The *Lilac Twig* and the *Flower Field* in (c) and (d) are results of the L-systems explained in Appendix C.

The following 8 plates are illustrations of complex dynamics discussed in Chapter 3.

Plate 17 $\pi \cos z$. The basin of attraction of the attracting fixed point at $-\pi$.

Plate 18 $2.97 \cos z$. The basin about to explode.

Plate 19 $2.96 \cos z$. After the explosion. Note the small black regions representing components of the complement of the Julia set.

Plate 20 $2.945 \cos z$. Note how the black regions have moved as the parameter is varied.

Plate 21 $2.935 \cos z$. A magnification shows that there is structure within the black regions that is reminiscent of the original picture.

Plate 22 $2.92 \cos z$. Two components of the stable sets about to collide at the point $-\pi$.

Plate 23 $2.919 \cos z$. Formation of smaller black regions within a component.

Plate 24 $2.918 \cos z$. A saddle-node periodic point about to bifurcate.

Plate 25 A 3D rendering of the Continuous Potential Method (CPM/M). A version of the cover of "The Beauty of Fractals" [83]. The "moon" is a Riemann sphere, which carries the structure of a Julia set.

Plate 26 A 3D rendering of the Continuous Potential Method (CPM/M), (cf. Figure 4.21(4)). The "sky" was obtained by the Level Set Method (LSM/M) close to the boundary of the Mandelbrot set and is displayed using pixel zoom.

Plate 27 A 3D rendering of the Continuous Potential Method (CPM/J) showing a filled-in Julia set in red.

Plate 28 A 2D rendering of the Continuous Potential Method (CPM/M) (cf. Figure 4.22).

Plate 29 A 2D rendering of the Distance Estimator Method (DEM/M). Points of same color have same distance to Mandelbrot set.

Plate 30 A 3D-rendering of the Distance Estimator Method (DEM/M). Points of same height and color have same distance to Mandelbrot set.

Plate 31 Diffusion Limited Aggregation (DLA) about a single point.

Plate 32 Four views of the three-dimensional fern of Figure 5.14 . The IFS code consists of four three-dimensional affine transformations.

Plate 33 Landscape with chimneys and smoke. This figure and Figure 34 computed from the same IFS database; which consists of 57 affine transformations.

Plate 34 Zoom on smoke shown in Plate 33.

Plate 35 A 3-dimensional cross section of $M_+ \cup M_-$ where $\text{Im}^b = 0$, see Section 4.3.4.

Plate 36 Fractal setting.

Captions for the cover images :

Plate 37 (Front cover) "Lake Mandelbrot". The Continuous Potential Method is applied to a section of the Mandelbrot set and rendered as a height field. Coloring and an added random fractal sky enhance "realism".

Plate 38 (Back cover top left) "Black Forest in Winter", several composed images using Iterated Function Systems (IFS) .

Plate 39 (Back cover top right) 3D rendering of Distance Estimator Method (DEM/M) for a blow up at the boundary of the Mandelbrot set. Points of same height have same distance to Mandelbrot set.

Plate 40 (Back cover center left) The floating island of *Gulliver's Travels* whose mathematically obsessed occupants lost all contact with the world around them. Shown here in the center of the zenith view of a fractal landscape in a lake opening onto fractal clouds (in homage to Magritte).

Plate 41 (Back cover center right) A foggy fractally cratered landscape. Light scattering with varying intensity through the same $T(x,y,z)$ as Figure 1.6(c) produces a fractal cloud with its shadows on the fractal landscape of Figure 1.5.

Plate 42 (Back cover bottom) Fractal landscape.

Credits for the color images

Besides the authors of the book several others have contributed to the pictures displayed on these pages and on the front and back cover of the book, which we gratefully acknowledge. Here is a list of the contributors and their images :

James Hanan, University of Regina
Arnaud Jacquin, Georgia Institute of Technology
Hartmut Jürgens, Universität Bremen
John P. Lewis, New York Institute of Technology
Francois Malassenet, Georgia Institute of Technology
John F. Mareda, Sandia National Laboratories
Gary A. Mastin, Sandia National Laboratories
Ken Musgrave, Yale University
Przemyslaw Prusinkiewicz, University of Regina
Laurie Reuter, George Washington University
Peter H. Richter, Universität Bremen
Marie-Theres Roeckerath-Ries, Universität Bremen
Alan D. Sloan, Georgia Institute of Technology
Peter A. Watterberg, Savannah River Laboratory

Plates 1 – 13	R.F. Voss
Plates 14 – 15	D. Saupe
Plate 16a	G.A. Mastin, P.A. Watterberg and J.F. Mareda
Plate 16b	J.P. Lewis
Plate 16c	J. Hanan and P. Prusinkiewicz
Plate 16d	P. Prusinkiewicz
Plates 17 – 24	R.L. Devaney
Plates 25 – 30	H. Jürgens, H.-O. Peitgen and D. Saupe
Plates 31	R.F. Voss
Plates 32 – 34	M.F. Barnsley, L. Reuter and A.D. Sloan
Plate 35	P. Richter and M.-T. Roeckerath-Ries
Plate 36	R.F. Voss
Front cover	H. Jürgens, H.-O. Peitgen and D. Saupe
Back cover top left	M.F. Barnsley, A. Jacquin, F. Malassenet, A.D. Sloan and L. Reuter
Back cover top right	H. Jürgens, H.-O. Peitgen and D. Saupe
Back cover center left	R.F. Voss
Back cover center right	R.F. Voss
Back cover bottom	B.B. Mandelbrot and F.K. Musgrave

there is detail at all scales, is carried out up to the screen resolution. The floating horizon method [92] is usually employed to render some coarse elevation data in a perspective view as in Figures 2.11 to 2.13 and 2.17 to 2.19. There each point is connected to its visible neighbors by a short line. In our case, where we have many more points, these lines generally can be omitted if each point is projected to one pixel on the screen and drawn. The hidden surface problem is solved most easily when the view direction is parallel to one coordinate plane. The farthest row of the data is drawn first, then the second last row, etc. Occasionally it may occur that two consecutive projected rows leave some pixels undefined in between them. For these pixels which are very easy to detect we must set some interpolated intermediate color and intensity. Most of the fascinating pictures of fractal terrain in Chapter 1 and of the potential surfaces for the Mandelbrot and Julia sets in Chapter 4 have been produced by such an algorithm (see the color plates). In the following we present a description of our extended version of the floating horizon method. It is sufficiently detailed as a guide for an implementation. Related algorithms are used in [38] and [41]. The method which is outlined below and refinements such as shadows and antialiasing will be the topic of our forthcoming paper [57].

2.7.3   The data and the projection

Let us assume for simplicity that the output of a fractal generator or of some other surface computation is presented as a square array of N by N elevation points. If

$$h : [0,1] \times [0,1] \rightarrow [0,1]$$

denotes a corresponding height function we may think of the data as a sample of h, namely

$$z_{i,j} = h(x_i, y_j)$$

where

$$x_i = \frac{i}{N-1}, y_j = \frac{j}{N-1}$$

and $i, j = 0, ..., N-1$. The surface to be rendered therefore is assumed to lie in the unit cube $[0,1]^3$. The projection of the data from the unit cube to the screen coordinates is most easily done by parallel projection such that one column of data elements is projected to one column on the screen. If the surface is to be viewed at an angle ϕ then the projection is given by

$$(x, y, z) \rightarrow (x, y \sin \phi + z \cos \phi).$$

Here $\phi = 0$ corresponds to a side view and $\phi = \frac{\pi}{2}$ yields a top view.

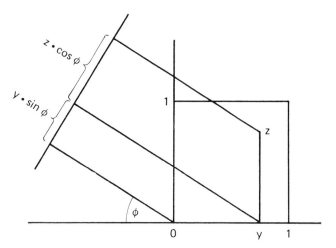

Fig. 2.20: Projection of data from unit cube to screen coordinates

2.7.4   A simple illumination model

For a local lighting model surface normals are essential. An approximate surface normal at the point

$$\vec{p} \;=\; (x_i, y_j, z_{i,j})$$

is given by

$$\vec{n} = (\,\frac{dx}{r}, \frac{dy}{r}, \frac{dz}{r}\,)$$

where

$$
\begin{aligned}
dx &= z_{i,j} - z_{i+1,j} \\
dy &= z_{i,j-1} - z_{i,j} \\
dz &= \frac{1}{N-1}
\end{aligned}
$$

and

$$r = \sqrt{dx^2 + dy^2 + dz^2}.$$

This will work very well for smooth surfaces. For fractal surfaces averages over several such approximations for nearby points are more suitable.

Let $\vec{l}$ denote the unit length vector based at $\vec{p}$ and pointing towards a light source. Let further $\vec{r}$ be the reflected light vector, i.e.

$$\vec{r} = 2 \cos \theta \cdot \vec{n} - \vec{l}$$

where θ is the angle enclosed between $\vec{n}$ and $\vec{l}$. If α denotes the angle between the view vector

$$\vec{v} = (\,0, -\cos \phi, \sin \phi\,)$$

and the reflected vector $\vec{r}$, then a simple illumination model, which takes into account only ambient, diffuse and specularly reflected light, may be written as

$$I(\vec{p}) = \begin{cases} I_a, & \text{if} \quad \cos\theta < 0 \\ I_a + f_d\cos\theta + f_s\cdot(\cos\alpha)^b, & \text{otherwise} \end{cases} \qquad (2.19)$$

Here I_a denotes the contribution of the ambient light, f_d and f_s are factors to be chosen as weights for the diffuse and specular components, and b is an exponent which controls the shininess of the surface. Satisfactory results can be obtained e.g. with

$$\begin{aligned} I_a &= 0.0 , \\ f_d &= 0.6 , \\ f_s &= 0.4 , \\ b &= 2.0 . \end{aligned}$$

For completeness let us remark that if the angle between the surface normal $\vec{n}$ and the view vector $\vec{v}$ is greater than $\frac{\pi}{2}$ in magnitude, then we are looking at the back side of the surface, and we can set the intensity to 0. In our algorithm *DisplayHeightField()* this case may only occur for the points in the first row. This simple model is a very crude one. For example, it ignores colors of the light and the surface. Better models are described in more detail in standard textbooks such as [39] or [92].

2.7.5   The rendering

Slices of the surface, which are parallel to the yz-plane, are projected onto a line on the screen, which is parallel to the screen y-axis. Therfore we can render these lines one by one and independently of each other. Define

$$P(y,z) = int\left[(N-1)(y\sin\phi + z\cos\phi)\right]$$

as the integer part of the y-value of the projection of a point (x,y,z) in pixel units. Then the algorithm for the rendering of all N columns of data looks as follows (see *DisplayHeightField()*). Here *Intens(i, k)* denotes the intensity calculated for the pixel *(i, k)* of the screen, while *I(x, y, z)* denotes the corresponding intensity for a point (x,y,z) on the surface.

   In the first loop all intensities in the i-th column are initialized to 0. Then the data points in the i-th column are projected one by one, starting at the front and moving to the back. The variable *horizon* keeps track of the current maximum of all projected y-values, and this explains the names of both the variable and

```
ALGORITHM DisplayHeightField (Z, N, phi)
Title         Computer graphics for surfaces given by array of elevation

Arguments    Z[][]         doubly indexed square array of heights
             N             number of elements in each row/column of Z
             phi           view angle
Globals      Intens[][]    output array for intensities of size at least N by √2 N
Variables    i, j, k       integers
             p, p0, p1     projected heights, screen y-coordinates
             x[]           N real x-values
             y[]           N real y-values
             horizon       integer (the floating horizon)
             h             real parameter for interpolation
Functions    P(y,z) = int [(N-1) * (y * sin(phi) + z * cos(phi)]
             I(x,y,z) = intensity function, see Eqn. (2.19)

BEGIN
    FOR i = 0 TO N-1 DO
        x[i] := i / (N - 1)
        y[i] := i / (N - 1)
    END FOR
    FOR i = 0 TO N-1 DO
        FOR k = 0, 1, 2, ... DO
            Intens[i][k] := 0
        END FOR
        p0 := P(y[0],z[i][0])
        Intens[0][p0] := I(x[i],y[0],z[i][0])
        horizon := p0
        FOR j = 1 TO N-1 DO
            p1 := P(y[j],z[i][j])
            IF p1 > horizon THEN
                Intens[i][p1] := I (x[i],y[j],z[i][j])
                p := p1 - 1
                WHILE p > horizon DO
                    h := (p - p0) / (p1 - p0)
                    Intens[i][p] := (1-h) * I(x[i],y[j-1],z[i][j-1]) + h * Intens[i][p1]
                    p := p - 1
                END WHILE
                horizon := p1
            END IF
            p0 := p1
        END FOR
    END FOR
END
```

the rendering method. If a point is projected below this horizon, it is not visible, and the normal and intensity calculations can be saved. Otherwise, the correct intensity is entered at the projected screen coordinates and the value of the horizon is updated. A special case arises when there are one or more pixels left out between the old and the new horizon. At these locations an interpolated intensity has to be applied. In the above scheme linear interpolation is employed, which corresponds to Gouraud shading. Alternatively one could interpolate surface normals and recompute intensities for the intermediate points, a procedure known as Phong shading.

It depends very much on the data and the projection angle ϕ chosen, which portion of the screen coordinate system actually should be displayed. In the worst case ($\phi = \frac{\pi}{4}$ and data occurring at $(x, 0, 0)$ as well as at $(x, 1, 1)$) only a viewport containing at least N by $\sqrt{2}\,N$ pixels will contain the whole surface. To get aesthetically pleasing results one obviously has to experiment with the projection and introduce proper scalings in the x and y screen coordinates.

2.7.6 Data manipulation

An alternative way to match the viewport to the diplayed object is to change the data instead. For example, a linear transformation

$$z_{i,j} = a \cdot h(x_i, y_j) + b$$

will shift the picture up or down and may scale the heights by the factor a. This may be desirable for other reasons as well. Often the height variations in the data are too strong or not pronounced enough for a nice picture. A linear scaling may help in this case. But also nonlinear scalings are of interest as color Plates 8 and 10 ($D = 2.15$ cubed height) demonstrates. In the case of the electrostatic potentials of the Mandelbrot set in Chapter 4 only a nonlinear scaling of the potential was able to change the initially rather dull looking images into something dramatic.

Another useful manipulation may be to pull down the heights of points near the front and/or the back of the surface in order to eliminate the impression of a surface that has been artificially cut off in the front and the back.

2.7.7 Color, anti-aliasing and shadows

The surface may have different colors at different places. E.g. color may be linked to height and possibly also to surface normals as seen in the color plates. It is easy to add color to our little rendering system just by multiplying the color of each projected point with the intensity as calculated in the algorithm. However, it must be noted, that this is straight forward only in the case that a full color (24 bit) graphics system is eventually used for display. Some restrictions apply for smaller systems with color lookup tables.

The algorithm *DisplayHeightField( )* can easily be extended to include anti-aliasing by means of supersampling. The principle is to first interpolate (multilinear or with splines) the surface to obtain a representation in a higher resolution. Then the algorithm is applied to the "inflated" surface and finally pixel averaging results in the anti-aliased image. This process can take advantage of the piecewise linear structure in the interpolated surface, thus, the computational burden of anti-aliasing is not as severe as it seems. In addition, distance fading as in [38] or simulated fog as in the color plates may be employed to further enhance picture quality.

The same algorithm that eliminates the hidden surfaces in our rendering may be used to calculate shadows from several light sources. This is the case because a shadow reflects nothing but the visibility from a given light source. Due to the simple structure of the algorithm only infinitely far light sources are admissible. In addition the vectors pointing to the light sources must be parallel to the xz-plane or the yz-plane. A complication arises : The rendering of the different data columns is no longer independent. Points in one column may cast a shadow on points of other columns. Thus, one complete row of data has to be rendered simultaneously. The cover picture was rendered with two light sources, one from behind the view point and the other from the right. The shadows from the second light source are clearly visible.

2.7.8 Data storage considerations

The storage requirements for the data may be quite stringent. For example, a common file may contain 1024 by 1024 floating point numbers, which typically amount to 4 mega bytes of storage. To reduce the size of these large files, several techniques may be of value. The first obvious one is to store the programs that generate the data in place of the data itself. This is a good idea for fractal surfaces generated by one of the cheaper methods such as midpoint displacement. However, for potentials of the Mandelbrot set and related images it is not advisable since the generation of a single set of data may take too long to be done twice or more often for several renderings.

To reduce the size of the data files we first note that we do not need the full range of precision and also of the exponents offered by floating point numbers. Thus, one elevation point may be conveniently stored as an unsigned 16 bit integer. This will reduce the above mentioned typical 4 mega bytes to only 2 mega bytes per file without decreasing the picture quality. But this still seems to be too much if one intends to keep many data files on hand.

The next step for the data reduction exploits the continuity of the height function or even its differentiability, if applicable as in the 3D renderings of Chapter 4. Instead of storing the absolute values of the height function we may, for each row j of data, store only the first elevation $z_{0,j}$ and the differences

$$\Delta^1_{i,j} = z_{i,j} - z_{i-1,j} \quad \text{for} \quad i = 1, ..., N-1 \; .$$

These first differences will be much smaller than the values $z_{i,j}$ thus, they may be stored in fewer bits per number. Still the original data may be reconstructed via

$$z_{i,j} = z_{i-1,j} + \Delta^1_{i,j} \quad \text{for} \quad i = 1, ..., N-1 \; .$$

If the height function h is also differentiable then the second differences

$$\Delta^2_{i,j} = \Delta^1_{i,j} - \Delta^1_{i-1,j} \quad \text{for} \quad i = 2, ..., N-1$$

or, more explicitly

$$\Delta^2_{i,j} = z_{i,j} - 2 z_{i-1,j} + z_{i-2,j} \quad \text{for} \quad i = 2, ..., N-1$$

can be expected to be even smaller in magnitude than the first differences. Thus, we may be able to store the numbers

$$z_{0,j}, \Delta^1_{1,j}, \Delta^2_{2,j}, \Delta^2_{3,j}, ..., \Delta^2_{N-1,j}$$

even more efficiently. The reconstruction of the j-th row of the original data then is achieved with one or two additions per point via

$$z_{1,j} = z_{0,j} + \Delta^1_{1,j}$$

and

$$\begin{aligned} \Delta^1_{i,j} &= \Delta^1_{i-1,j} + \Delta^2_{i,j} \\ z_{i,j} &= z_{i-1,j} + \Delta^1_{i,j} \end{aligned}$$

for $i = 2, ..., N-1$. For many of the smooth surfaces of Chapter 4 this technique resulted in a requirement of only roughly 8 bits per data element thus reducing the initial 4 mega bytes to only about one mega byte.

There is even more structure in the first and second differences which can be used to further reduce the size of the data files. For this purpose data compression techniques such as simple run length encoding, or more advanced methods such as Huffman encoding [89] or Lempel-Ziv-Welch encoding [106] as implemented in the UNIX *compress* utility may result in an additional saving. Finally, one may modify these system compression techniques to better fit our case of interest.

2.8   Random variables and random functions

Random variables and random functions are the two major building blocks used in random fractals. Here we review some of the basic definitions (see e.g. [8,107]). [44] is another elementary text for probability theory.

Sample space. An abstract representation for a trial and its outcomes such as one throw of a die or ideally also one call to a (pseudo) random number generator is the *sample space*. Each distinguishable and indecomposable outcome, or simple event, is regarded as a point in sample space S. Thus, for a typical random number generator the sample space would consist of all integers between 0 and a very large number such as $2^{31} - 1$. Every collection of simple events i.e. a set of points in S is called an *event*.

Probability. For every event E in the sample space S, i.e. $E \subset S$ we assign a non-negative number, called the *probability* of E denoted by $P(E)$ so that the following axioms are satisfied:

(a) for every event E, $P(E) \geq 0$

(b) for the certain event, $P(S) = 1$

(c) for mutually exclusive events E_1, E_2, $P(E_1 \cup E_2) = P(E_1) + P(E_2)$

Random variables. A *random variable* X is a function on a sample space S with two properties

(a) the values it assumes are real numbers,

(b) for every real number x the probability that the value of the function is less than or equal to x can be calculated, and is denoted by $P(X \leq x)$.

In the example of dice or random number generators the points in the sample space are themselves numerical outcomes, and the outcome is itself a random variable.

Probability density function. Suppose that X is a random variable and there is a function $f(x)$ such that

(a) $f(x) \geq 0$.

(b) $f(x)$ has at most a finite number of discontinuities in every finite interval on the real line.

(c) $\int_{-\infty}^{\infty} f(x) \, dx = 1$.

(d) For every interval $[a, b]$

$$P(a \leq X \leq b) = \int_a^b f(x)\,dx.$$

Then X is said to be a *continuous random variable* with *probability density function $f(x)$*.

Distribution function. A useful picture of the spread of probability in a distribution is provided by the *distribution function $F(x)$*, which is defined as

$$F(x) = P(X \leq x).$$

From its definition, $0 \leq F(x) \leq 1$, $F(-\infty) = 0$, $F(\infty) = 1$, $P(a \leq X \leq b) = F(b) - F(a) \geq 0$. For a continuous random variable with probability density $f(x)$ we have

$$F(x) = \int_{-\infty}^x f(t)\,dt.$$

Uniform distribution. A continuous random variable X is said to have a *uniform distribution* over an interval $[a, b]$, if its probability density function $f(x)$ satisfies

$$f(x) = \begin{cases} \frac{1}{b-a}, & \text{if } a \leq x \leq b \\ 0, & \text{otherwise} \end{cases}$$

If $a \leq x \leq b$, then

$$P(X \leq x) = \frac{x - a}{b - a}.$$

Normal distribution. A continuous random variable X is said to have a *normal distribution* with parameters μ and $\sigma > 0$ if its probability density function $f(x)$ is

$$f(x) = \frac{1}{\sqrt{2\pi}\sigma} e^{-\frac{1}{2}(\frac{x-\mu}{\sigma})^2}.$$

Such a random function is said to be distributed $N(\mu, \sigma)$. The graph of f has the familiar Gaussian bell shape centered around μ. The routine *Gauss()* in Section 2.2 is constructed to approximate a random variable with a distribution $N(0, 1)$. Its distribution function $F(x)$ is tabulated in most handbooks on statistics.

Expectation. If X is a continuous random variable with distribution function $f(x)$ then the *mean* or the *expectation* of X, written as $E(X)$ is defined as

$$E(X) = \int_{-\infty}^\infty x f(x)\,dx$$

provided that

$$\int_{-\infty}^\infty |x| f(x)\,dx$$

converges. Otherwise, $E(X)$ is said not to exist. The name has its justifica-
tion in the weak law of large numbers, which states, roughly speaking, that if
$X_1, X_2, \ldots, X_n$ is a random sample from a distribution for which $E(X)$ exists,
then the average

$$\overline{X} = \sum_{i=1}^{n} \frac{X_i}{n}$$

converges in probability to the mean $E(X)$. For the uniformly distributed random variable defined two paragraphs above we have of course $E(X) = \frac{a+b}{2}$,
whereas for a random variable with a normal distribution $N(\mu, \sigma)$ we obtain
$E(X) = \mu$. The expectation E is a linear function in the sense that for two
random variables X and Y and real numbers a and b we have $E(aX + bY) =
aE(X) + bE(Y)$. Moreover, if $g(x)$ is a real function, then $Y = g(X)$ is
another random variable and its expectation is

$$E(g(X)) = \int_{-\infty}^{\infty} g(x) f(x) \, dx.$$

Variance. Generally a random variable is not adequately described by just stat-
ing its expected value. We should also have some idea of how the values are
dispersed around the mean. One such measure is the *variance*, denoted by var X
and defined as

$$\operatorname{var} X = E[(X - E(X))^2].$$

The positive square root of the variance is known as the *standard deviation* of
X, and is denoted by σ. We also have

$$\operatorname{var} X = E(X^2) - E(X)^2.$$

The variance of a random variable which is uniformly distributed over $[a, b]$ is
easily calculated as

$$\operatorname{var} X = \frac{1}{12}(b - a)^2$$

and for random variables with normal distribution $N(\mu, \sigma)$ the variance is σ^2.
If two random variables X, Y are independent, i.e.

$$P(X \leq x \text{ and } Y \leq y) = P(X \leq x) + P(Y \leq y)$$

for all $x, y \in \mathbf{R}$, then we have for all $a, b \in \mathbf{R}$

$$\operatorname{var}(aX + bY) = a^2 \operatorname{var} X + b^2 \operatorname{var} Y.$$

Covariance and correlation. A measure which associates two random variables is given by the *covariance*. The covariance of X, Y, written $C(X, Y)$, is
defined as

$$C(X, Y) = E[(X - E(X))(Y - E(Y))].$$

If the two random variables are independent, then their covariance is 0, of course. If units of X or Y are changed then the covariance also changes. This defect is removed by considering the *correlation coefficient*, which is the covariance divided by both standard deviations.

Random functions. Let T be a subset of the real line. By a *random function* of an argument $t \in T$ we mean a function $X(t)$ whose values are random variables. Thus, a random function X is a family of random variables $X(s)$, $X(t)$, etc. Formally, a random function is considered to be specified if for each element $t \in T$ we are given the distribution function

$$F_t(x) = P(X(t) \leq x)$$

of the random variable $X(t)$, and if also for each pair of elements $t_1, t_2 \in T$ we are given the distribution function

$$F_{t_1,t_2}(x_1, x_2) = P(X(t_1) \leq x_1 \text{ and } X(t_2) \leq x_2)$$

of the two-dimensional random variable $Z(t_1, t_2) = (X(t_1), X(t_2))$, and so on for all finite dimensional random variables

$$Z(t_1, t_2, \ldots, t_n) = (X(t_1), X(t_2), \ldots, X(t_n))$$

Stationarity. The random function will be called *stationary*, if all the finite dimensional distribution functions defined above remain the same when all points $t_1, t_2, \ldots, t_n$ are shifted along the time axis, i.e. if

$$F_{t_1+s,t_2+s,\ldots,t_n+s}(x_1, x_2, \ldots, x_n) = F_{t_1,t_2,\ldots,t_n}(x_1, x_2, \ldots, x_n)$$

for any $t_1, t_2, \ldots, t_n$ and s. The physical meaning of the concept of stationarity is that $X(t)$ describes the time variation of a numerical characteristic of an event such that none of the observed macroscopic factors influencing the occurrence of the event change in time.

Stationary increments. Fractional Brownian motion is not a stationary process, but its increments

$$X(t + s) - X(t)$$

are. In general, we say that the random function $X(t)$ has *stationary increments*, if the expected value of the increment is 0

$$E[X(t + s) - X(t)] = 0$$

and if the *mean square increments*

$$E[|X(t + s) - X(t)|^2]$$

do not depend on time t.

Chapter 3

Fractal patterns arising in chaotic dynamical systems

Robert L.Devaney

3.1   Introduction

Fractals are everywhere. This is no accident, because even the simplest mathematical expressions, when interpreted as dynamical systems, yield fractals. The goal of this chapter is to make this statement precise. We will describe some of the elements of the field of mathematics known as dynamical systems and show how fractals arise quite naturally in this context. We will also present some of the algorithms by which some of these fractal patterns may be generated.

This chapter is aimed at a circle of readers of non-mathematicians, although it must be said that some mathematical sophistication is helpful in many spots. Many of the topics treated here are of current research interest in mathematics. In fact, some of the dynamical behavior we will encounter has not yet been explained completely. Despite this, the formulation of these problems is quite straightforward and experimental work related to these problems is accessible to anyone with a good high school background in mathematics and access to a computer.

Our approach to these problems will be a combination of mathematical techniques and computer graphics. As the reader is undoubtedly aware, many of the fractal images generated by dynamical systems are both interesting and alluring. Moreover, the fact that these images are generated by such simple systems as $z^2 + c$ or e^z or $\sin z$ make this field quite accessible to scientists in all disciplines. Finally, as the reader will see, the mathematics itself is quite beautiful. All in all, this field of mathematics offers quite a lot to scientists: aesthetic appeal, accessibility, and applicability. We hope that this chapter convey some of these elements to the reader.

3.1.1 Dynamical systems

What is a dynamical system? Basically, any process which evolves in time is an example of a dynamical system. Such systems occur in all branches of science and, indeed, in virtually every aspect of life itself. Weather patterns are examples of huge dynamical systems: the temperature, barometric pressure, wind direction and speed and amount of precipitation are all variables which change in time in this system. The economy is another example of a dynamical system: the rise and fall of the Dow Jones average is a simple illustration of how this system fluctuates in time. The evolution of the planets in the solar system and simple chemical reactions are examples of other dynamical systems.

The field of dynamical systems deals with these and many other types of physical and mathematical processes. The basic goal of this field is to predict the eventual outcome of the evolving process. That is, if we know in complete detail the past history of a process that is evolving in time, can we predict what will happen to it in the future? In particular, can we deduce the long term or asymptotic behavior of the system?

The answer to this question is sometimes yes and sometimes no. Obviously, prediction of weather systems and stock market fluctuations cannot be made in the long term. On the other hand, we are all confident that the sun will rise tomorrow morning and that no extraordinary chemical reaction will take place when we add cream to our coffee at that time. So some dynamical systems are predictable while others are not.

What makes some dynamical systems predictable and others unpredictable? From the above examples, it would seem that dynamical systems which involve a huge number of variables like the weather systems or the economy are unpredictable, whereas systems with only a few variables are easier to understand. However, while this may be true in some cases, it is by no means true in general. Even the simplest of dynamical systems depending upon only one variable may yield highly unpredictable and essentially random behavior. The reason for this is the mathematical notion of chaos, which may pervade even the simplest of dynamical systems. This is one of the major achievements of mathematics in the last few years, the recognition that simple, deterministic systems may behave unpredictably or randomly.

In the succeeding sections, we will give several mathematical examples of chaotic systems. All will have one trait in common: dependence upon only one or two variables. And many, when looked at geometrically using computer graphics, will exhibit fractals as the underlying geometric feature.

3.1.2 An example from ecology

Let us illustrate how systems may behave unpredictably by describing a simple mathematical dynamical system which has its roots in ecology.

Suppose there is a single species whose population grows and dwindles over time in a controlled environment. Ecologists have suggested a number of mathematical models to predict the long-time behavior of this population. Here is one of the simplest of their models.

Suppose we measure the population of the species at the end of each generation. Rather than produce the actual count of individuals present in the colony, suppose we measure instead the percentage of some limiting number or maximum population. That is, let us write P_n for the percentage of population after generation n, where $0 \leq P_n \leq 1$. One simple rule which an ecologist may use to model the growth of this population is the logistic equation

$$P_{n+1} = kP_n(1 - P_n).$$

where k is some constant that depends on ecological conditions such as the amount of food present. Using this formula, the population in the succeeding generation may be deduced from a knowledge of only the population in the preceding generation and the constant k.

Note how trivial this formula is. It is a simple quadratic formula in the variable P_n. Given P_n and k, we can compute P_{n+1} exactly. In Table 3.1 we have listed the populations predicted by this model for various values of k. Note several things. When k is small, the fate of the population seems quite predictable. Indeed, for $k = 0.5$, the population dies out, whereas for $k = 1.2$, 2, and 2.7, it tends to stabilize or reach a definite limiting value. Above 3, different values of k yield startlingly different results. For $k = 3.1$, the limiting values tend to oscillate between two distinct values. For $k = 3.4$, the limiting values oscillate between four values. And finally, for $k = 4$, there is no apparent pattern to be discerned. One initial value, $P_0 = 0.5$, leads to the disappearance of the species after only two generations, whereas $P_0 = 0.4$ leads to a population count that seems to be completely random.

$$P_{n+1} = kP_n(1 - P_n)$$

				k			
0.5	1.2	2.0	2.7	3.1	3.4	4.0	4.0
.5	.5	.5	.500	.5	.5	.4	.5
.125	.3	.5	.750	.775	.85	.96	1
.055	.252	.5	.563	.540	.434	.154	0
.026	.226	.5	.738	.770	.835	.520	0
.013	.210	.5	.580	.549	.469	.998	0
.006	.199	.5	.731	.768	.847	.006	0
.003	.191	.5	.590	.553	.441	.025	0
.002	.186	.5	.726	.766	.838	.099	0
.001	.181	.5	.597	.555	.461	.358	0
.000	.178	.5	.722	.766	.845	.919	0
.000	.176	.5	.603	.556	.446	.298	0
.000	.174	.5	.718	.765	.840	.837	0
.000	.172	.5	.607	.557	.457	.547	0
.000	.171	.5	.716	.765	.844	.991	0
.000	.170	.5	.610	.557	.448	.035	0
.000	.170	.5	.713	.765	.841	.135	0
.000	.169	.5	.613	.557	.455	.466	0
.000	.168	.5	.711	.765	.843	.996	0
.000	.168	.5	.616	.557	.450	.018	0
.000	.168	.5	.710	.765	.851	.070	0
.000	.168	.5	.618	.557	.455	.261	0
.000	.168	.5	.708	.765	.843	.773	0

Table 3.1: Values of P_n for various k-values

This is the unpredictable nature of this process. Certain k-values lead to results which are quite predictable – a fixed or periodically repeating limiting value. But other k-values lead to results which are, for all intent and purposes, random.

The reader may object that the quite limited table of values for P_n when $k =$ 4 can in no way be interpreted as a proof that the values do behave randomly. Nevertheless, it is a true fact which may be proved quite easily with techniques from dynamical systems theory. See, for example, [25].

3.1.3 Iteration

The principal ingredient in the mathematical formulation of the example from ecology was iteration. Iteration means repeating a process over and over again. That is just what we did to generate the list of successive population values for the logistic equation. Let us formulate this process in a slightly different manner. Consider the mathematical function $F(x) = kx(1 - x)$. This function assigns to each number x a new number, the image of x, which we call $F(x)$. So, as is usual with mathematical functions, F gives a rule for converting x to a new value, in this case $F(x)$, or $kx(1 - x)$.

We may repeat this process by computing $F(F(x))$. This means that we compute

$$
\begin{aligned}
F(F(x)) &= F(kx(1 - x)) \\
&= k[kx(1 - x)][1 - kx(1 - x)]
\end{aligned}
$$

and we can do it again, finding

$$F(F(F(x))) = F(k[kx(1 - x)][1 - kx(1 - x)])$$

And so forth. If x represents our initial population P_0 from the previous example, then we have

$$
\begin{aligned}
P_1 &= F(x), \\
P_2 &= F(F(x)), \\
P_3 &= F(F(F(x)))
\end{aligned}
$$

and so forth. To make the notation less cumbersome, let us introduce the notation $F^n(x)$ to mean the n-fold iteration of F; that is,

$$
\begin{aligned}
F^1(x) &= F(x), \\
F^2(x) &= F(F(x)), \\
F^3(x) &= F(F(F(x)))
\end{aligned}
$$

Note that $F^2(x)$ does *not* mean $(F(x))^2 = F(x) \cdot F(x)$. Rather, $F^2(x)$ is the second iterate of the expression F.

Iteration is a very simple process that is perhaps easiest to explain using a scientific calculator. Consider such a calculator with its many special function keys like "x^2", "$\sqrt{x}$", "$\sin x$", "$\exp x$", and so forth. Each of these represents a mathematical function. Suppose we input a particular number and then repeatedly strike the "$\sqrt{x}$" key. What are we doing? We are simply iterating the square root function. For example, if we let $S(x) = \sqrt{x}$ and input $x = 256$ into the calculator, then repeated strikes of "$\sqrt{x}$" yields the following iterates of 256:

$$
\begin{aligned}
S(256) &= 16 \\
S^2(256) &= 4 \\
S^3(256) &= 2 \\
S^4(256) &= 1.414214\ldots \\
S^5(256) &= 1.189207\ldots \\
S^6(256) &= 1.090508\ldots \\
S^7(256) &= 1.044274\ldots \\
&\vdots \\
S^{20}(256) &= 1.000005\ldots
\end{aligned}
$$

Notice that successive iterates of S started with 256 converge quite quickly to 1. In fact, this happens no matter which number is initially input into the calculator. For example, if the initial input is 200, we get

$$
\begin{aligned}
S(200) &= 14.14214\ldots \\
S^2(200) &= 3.760603\ldots \\
S^3(200) &= 1.939227\ldots \\
S^4(200) &= 1.392561\ldots \\
&\vdots \\
S^{20}(200) &= 1.000005\ldots
\end{aligned}
$$

Even if the initial input x satisfies $0 < x < 1$, we still find that iterates tend to 1:

$$
\begin{aligned}
S(0.2) &= 0.4472136\ldots \\
S^2(0.2) &= 0.6687403\ldots \\
S^3(0.2) &= 0.8177654\ldots \\
&\vdots \\
S^{20}(.2) &= 0.9999985\ldots
\end{aligned}
$$

Let's iterate more of these functions. For example, what happens if we repeatedly strike the "x^2" key. Clearly, if $x > 1$, repeated squaring tends to enlarge the results. In fact, after only a few iterations, repeated squaring leads to an overflow message from the computer. If we write $T(x) = x^2$, another way of saying this is

$$T^n(x) \to \infty \text{ as } n \to \infty \quad \text{if} \quad x > 1.$$

The nth iterate of any $x > 1$ tends to get arbitrarily large if $x > 1$. What if $0 < x < 1$? Then iteration of T yields a different answer. Successive squarings of such numbers yield smaller and smaller results, so

$$T^n(x) \to 0 \text{ as } n \to \infty \quad \text{when} \quad 0 < x < 1.$$

Finally, in the intermediate case, $x = 1$, it is clear that $T^n(x) = 1$ for all n.

So the iteration of T yields three different behaviors, depending upon whether $0 < x < 1$, $x = 1$, or $x > 1$. You can easily extend this to the case of negative x-values. Notice one thing: unlike iteration of $F(x) = kx(1 - x)$, you can easily predict the fate of any x under iteration of T.

3.1.4 Orbits

The list of successive iterates of a point or number is called the *orbit* of that point. We use the calculator to compute some other orbits. For example, let us use the "sin x" button on the calculator to compute the orbit (in radians) of any initial input. If we let $S(x) = \sin x$ and choose the initial value $x = 123$, we find

$$
\begin{aligned}
S(123) &= -0.459\ldots \\
S^2(123) &= -0.443\ldots \\
S^3(123) &= -0.429\ldots \\
S^4(123) &= -0.416\ldots \\
&\vdots \\
S^{17}(123) &= -0.312\ldots \\
S^{18}(123) &= -0.307\ldots \\
S^{19}(123) &= -0.302\ldots \\
S^{20}(123) &= -0.298\ldots
\end{aligned}
$$

Slowly, ever so slowly, successive iterates of sin x tend to 0:

$$
\begin{aligned}
S^{73}(123) &= -0.185\ldots \\
S^{74}(123) &= -0.184\ldots \\
S^{75}(123) &= -0.183\ldots
\end{aligned}
$$

$$\vdots$$

$$S^{148}(123) = -0.135\ldots$$
$$S^{149}(123) = -0.135\ldots$$
$$S^{150}(123) = -0.134\ldots$$

$$\vdots$$

$$S^{298}(123) = -0.098049\ldots$$
$$S^{299}(123) = -0.097892\ldots$$
$$S^{300}(123) = -0.097580\ldots$$

So the orbit of $x = 123$ is the sequence of numbers 123, -0.459..., -0.443..., -0.429... and we have

$$S^n(123) \to 0 \text{ as } n \to \infty$$

That is, the orbit of 123 tends asymptotically to 0. We leave it to the reader to check that this happens no matter what x is initially input. That is

$$S^n(x) \to 0 \text{ as } n \to \infty \text{ for all } x.$$

This brings us back to the basic question in dynamical systems: can we predict the fate of orbits under iteration. This is similar to the question we started with: can we predict the weather or can we predict the behavior of the Dow Jones average?

For all of the systems we have discussed using the calculator, namely "$\sqrt{x}$", "x^2", and "sin x", the answer has been yes. Here is one last example where the fate of orbits can be decided, but the result is not so easy to predict ahead of time. What happens when the cosine function is iterated? Let's see. Let $C(x) = \cos x$ and choose any input, say $x = 123$. Then, in radians, we find

$$C(123) = -0.887\ldots$$
$$C^2(123) = 0.630\ldots$$
$$C^3(123) = 0.807\ldots$$
$$C^4(123) = 0.691\ldots$$
$$C^5(123) = 0.770\ldots$$

$$\vdots$$

$$C^{99}(123) = 0.739085\ldots$$
$$C^{100}(123) = 0.739085\ldots$$
$$C^{101}(123) = 0.739085\ldots$$

So the orbit of 123 is a sequence of numbers that tends to 0.739085... Again, this happens no matter which x is initially input into the calculator, but this result could hardly have been guessed ahead of time.

Nevertheless, once you have seen a few such examples of iteration of cosine, you come quickly to the conclusion that all orbits tend to .739085... or, to introduce a technical term, all orbits of the cosine function are *stable*. More precisely, a stable orbit is one which has the property that, if you change the initial input slightly, the resulting orbit behaves more or less similarly. This happens for $S(x) = \sqrt{x}$, since all initial inputs lead to orbits which tend to 1. This happens for $T(x) = x^2$, as long as $0 < x \neq 1$, for orbits of $x < 1$ tend to 0 while orbits of $x > 1$ tend to ∞.

One should think of stable orbits as "good" orbits. We mean good here in the following sense. Suppose your dynamical system represents a physical process whose outcome you would like to predict. Now, in setting up the physical model, you will probably make small errors in your observations that lead to your choice of an initial input. If the resulting orbit is stable, then chances are that these small errors will not alter the ultimate behavior of your system, and so the predictions based on the model will be more or less accurate.

Another reason why stable orbits are good orbits arises when round-off errors are encountered. Generally, when a dynamical system is iterated on a computer or calculator, each successive iteration yields only an approximation of the next value. That is, the computed orbit may differ slightly at each stage from the actual orbit. It is true that these round-off errors may accumulate and lead to major errors in the predictions. But usually, if the initial orbit is stable, these round-off errors will not matter. Small changes introduced at each stage of the computation will not affect the ultimate behavior of the orbit.

3.2 Chaotic dynamical systems

3.2.1 Instability: The chaotic set

Unfortunately, not all orbits of dynamical systems are stable. It is an unpleasant fact of life that even the simplest of dynamical systems possess orbits which are far from being stable. We call these orbits *unstable*. An unstable orbit is one for which, arbitrarily close to the given initial input, there is another possible input whose orbit is vastly different from the original orbit.

Unstable orbits come in many different guises. The unpredictable orbits of $F(x) = 4x(1 - x)$, which seem to jump around randomly, are one type of unstable orbit. But there are other, simpler orbits which should also be termed unstable.

For example, consider again the squaring function

$$T(x) = x^2$$

The orbit of 1 is quite simple: $T(1) = 1$ so $T^n(1) = 1$ for all n. This is an example of a *fixed point* for the dynamical system. But this simple orbit is unstable. The reason is, suppose we make a slight error in computing the initial value 1. Then our input is $x \neq 1$, and, as we have already seen, if $x > 1$ then $T^n(x) \to \infty$ whereas if $0 < x < 1$, then $T^n(x) \to 0$. Thus, nearby initial conditions lead to vastly different types of orbits, some tending to 0 and others tending to ∞.

Clearly, it is of great importance to understand the set of all points in a given dynamical system whose orbits are unstable. Is the set of unstable orbits large, indicating a high degree of instability or unpredictability in the system? Or is it small, suggesting that the model we are using possesses a high probability of actually producing good results. Unfortunately, many simple dynamical systems possess large sets of initial conditions whose orbits are unstable. We will call the set of all points whose orbit is unstable the *chaotic set*. As we shall see, small changes of parameter may radically alter the makeup of the chaotic set. Nevertheless, to mathematicians, the important question is: what does this chaotic set look like? Is it large or small? Is it a simple set or a complicated one? And how does it vary as the dynamical system itself changes?

This is where fractals enter the field of dynamical systems. Very often, the set of points whose orbits are unstable form a fractal. So these fractals are given by a precise rule: they are simply the chaotic set of a dynamical system.

3.2.2   A chaotic set in the plane

Up to now, all of our examples of dynamical systems have been one dimensional. But most phenomena in nature depend upon more than one variable (often many more variables), so it is appropriate to study higher dimensional dynamical systems. For the remainder of this paper we will consider two dimensional dynamical systems. The reason for this is twofold. First, mathematically speaking, much of what occurs in one-dimensional dynamics is fairly well (but not completely!) understood. On the other hand, two-dimensional dynamics presents a completely different set of possibilities which, to this day, remain unsolved. Second, as can be imagined, two dimensional dynamics are readily studied using computer graphics, with its conveniently two-dimensional graphics screen.

Let us begin with a rather simple dynamical system known as the Hénon map , named for the French astronomer M. Hénon who first encountered this map in 1974. This dynamical system involves two variables (x_n, y_n) which evolve according to the rule

$$\begin{aligned} x_{n+1} &= 1 + y_n - Ax_n^2 \\ y_{n+1} &= Bx_n \end{aligned} \qquad (3.1)$$

Here A and B are parameters which we will fix at $A = 1.4$, $B = 0.3$. Many, many other phenomena can be observed for other A and B-values, but we will leave the investigation of these values to the reader. So we look at the dynamical system

$$\begin{aligned} x_{n+1} &= 1 + y_n - 1.4\,x_n^2 \\ y_{n+1} &= 0.3\,y_n \end{aligned} \qquad (3.2)$$

This dynamical system possesses a *strange attractor*, a mathematical object which, even today, defies mathematical analysis. Let us iterate the Hénon map using various different initial values (x_0, y_0) and also plotting the successive iterates graphically. Figure 3.1a shows the results when we plotted 100,000 iterations of the initial point (0,0). Figure 3.1b shows the same plot, but this time we chose a different initial condition $(-1, 0)$. In fact, if you experiment with a number of different initial conditions, you will find that either the orbit tends to the structure shown in Figure 3.1, or else the orbit tends to ∞. In fact, these are the only possibilities, as was proved by Hénon [51].

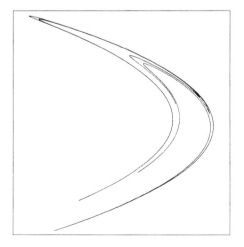 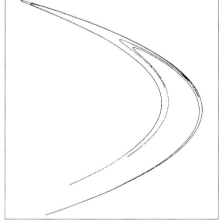

Fig. 3.1: Two plots of the Hénon attractor. The orbit of $(0, 0)$ is depicted in Figure 3.1a, and the orbit of $(-1, 0)$ is shown in Figure 3.1b . Horizontally x varies from -1.5 to 1.5, vertically y varies from -0.4 to 0.4.

So what is happening? Many orbits tend under iteration to land on the same set Λ as depicted in Figure 3.1 . Λ is called an *attractor* because all sufficiently nearby orbits converge to it. Λ is a *strange attractor* because it is not a "simple" object like a point or cycle of points (a "periodic orbit"). In fact, Λ has all of the attributes of a fractal, as demonstrated in Figure 3.2 . This figure depicts several successive enlargements which tend to separate bands in Λ, a typical occurrence in Cantor-like sets.

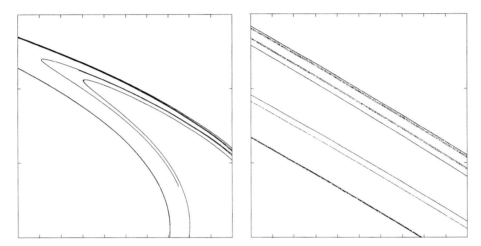

Fig. 3.2: Two successive enlargements of the Hénon attractor. The same 100,000 points as in Figure 3.1a are used. The regions shown are $0 \leq x \leq 1$ and $0 \leq y \leq 0.3$ on the left and $0.7 \leq x \leq 0.8$ and $0.15 \leq y \leq 0.18$ on the right.

We remark again that the exact structure of Λ is not understood and is an important open problem in mathematics. We also encourage the reader to experiment with the Hénon map with different parameter values, i.e., with different values for A and B. There is a vast collection of different fractal-like phenomena which occurs for various parameter values. This is interspersed with other parameter values which exhibit no such behavior.

So where is the chaos? The chaotic dynamics occur on the attractor Λ itself. Here is a simple experiment to see this. Take two initial conditions which are very close to each other and whose orbits tend to Λ. Then list the first 100 iterates of each point– write them out, don't plot them. You will see that, after just a very few iterations, the lists of successive points on the orbit have little to do with one another. The orbits are quite distinct. Nevertheless, if you now plot these orbits, you see that the exact same picture results. That is to say, nearby orbits both tend to Λ, but once they are there, they are hopelessly scrambled. So the strange attractor Λ forms our set of chaotic orbits.

3.2.3 A chaotic gingerbreadman

Here is another example of a chaotic set in the plane. Consider the dynamical system given by

$$
\begin{aligned}
x_{n+1} &= 1 - y_n + |x_n| \\
y_{n+1} &= x_n
\end{aligned}
\qquad (3.3)
$$

This is a piecewise linear mapping of the plane which is chaotic in certain regions and stable in others. For example, the point $(1,1)$ is a fixed point, and all other points within the hexagonal region H whose vertices are $(0,0)$, $(1,0)$, $(2,1)$, $(2,2)$, $(1,2)$, and $(0,1)$ are periodic with period 6. Indeed, this region lies in the right half plane where the dynamical system is the simple affine transformation

$$
\begin{aligned}
x_{n+1} &= 1 - y_n + x_n \\
y_{n+1} &= x_n
\end{aligned}
$$

which is a periodic rotation of period 6 about $(1,1)$. So the orbit of any point which remains in H (except $(1,1)$) is a finite set of 6 points.

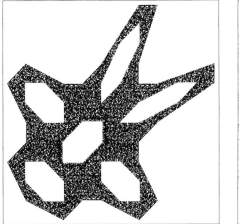 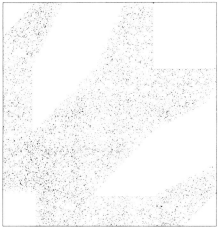

Fig. 3.3: The "gingerbreadman" and an enlargement.

This should be contrasted to the orbit of a point just outside of this region: Figure 3.3a displays the first 100,000 iterates of $(-0.1, 0)$. Note how this orbit "fills" a region that looks like a "gingerbreadman." There is the stable hexagonal region H forming the belly and five other regions forming the legs, arms, and head of the gingerbreadman. In these five regions there is a unique orbit of period 5, and all other points have period $6 \cdot 5 = 30$. These points are, of course, points with stable orbits. The orbit which gives the shaded region in Figure 3.3 apparently enters every small region except H and the period 5 hexagon. Indeed, it may be proved rigorously that there is such an orbit that wanders around the gingerbreadman, coming arbitrarily close to any preassigned point.

An enlargement of a portion of Figure 3.3a is shown in Figure 3.3b . Note how the chosen orbit is evenly distributed in this enlargement.

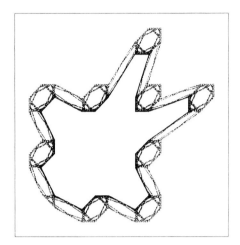 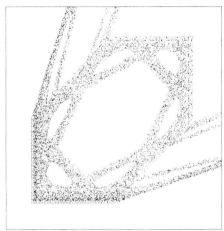

Fig. 3.4: Another "gingerbreadman" and enlargement.

One may conjecture that all orbits in the gingerbreadman have this property. Actually, this is by no means true. It can be proved that there are infinitely many distinct periodic orbits which have arbitrarily large period. In fact, in any region, no matter how small, there are infinitely many points which lie on periodic orbits! As a consequence, the entire gingerbreadman, minus the stable regions above, consists of unstable orbits.

It turns out that all points on the boundary of the gingerbreadman are periodic – these points form the polygonal "skin" of the gingerbreadman. But just on the other side of this polygon lies another chaotic region: see Figure 3.4 . In fact, there is a succession of chaotic regions, each separated by polygons from the others. As we see, chaotic regimes may fill up space, as in this example, as well as form fractals, as in the previous example. For more details and proofs of the above facts, we refer the reader to [24].

3.3   Complex dynamical systems

3.3.1   Complex maps

Now we come to one of the most alluring areas in dynamical systems, the study of the dynamics of complex maps. This study was initiated by P. Fatou and G. Julia in the early twentieth century. After a brief period of intense study during 1910-1925, the field all but disappeared. It was revived about ten years ago, thanks mainly to the discovery by Mandelbrot of the well known set that bears his name as well as to the spectacular computer graphics images that typically accompany these dynamical systems.

Here we will discuss only the Julia sets of complex maps from an introductory point of view. The related study of the Mandelbrot set, and more sophisticated algorithms to compute Julia sets will be discussed in the next chapter. Our goal is to describe the Julia sets in terms of the chaotic dynamics described above. We will also present the algorithms by which various Julia sets may be generated.

We begin with a simple dynamical system, $F(z) = z^2$. Here z is a complex number. That is, $z = x + iy$ where x and y are real numbers and i is the imaginary constant $\sqrt{-1}$. We operate algebraically with complex numbers just as with pairs of real numbers, so

$$
\begin{aligned}
z^2 = z \cdot z &= (x + iy) \cdot (x + iy) \\
&= x^2 + i^2 y^2 + 2ixy \\
&= x^2 - y^2 + i(2xy)
\end{aligned}
$$

Note that any complex number has two constituent pieces, the *real part* x and the *imaginary part* y. To specify a complex number, we must specify both the real and the imaginary part.

Since complex numbers are given by pairs of real numbers, they may be identified with points in the Cartesian plane in the obvious way, with the real part giving the x-coordinate and the imaginary part giving the y-coordinate.

We also define the *absolute value* of a complex number $z = x + iy$ by

$$
|z| = \sqrt{x^2 + y^2}
$$

That is, $|z|$ just measures the distance from the origin to z. This of course is exactly what the usual absolute value on the real line does.

Now let us consider what happens when we iterate $F(z) = z^2$. Recall that we investigated this exact dynamical system on the real line in Section 3.1.3 when we introduced the notion of iteration. Suppose we start with a point $z_0 = x_0 + iy_0$ with $|z_0| < 1$. Then

$$
\begin{aligned}
|F(z_0)| &= |x_0^2 - y_0^2 + i \cdot 2x_0 y_0| \\
&= \sqrt{x_0^4 + y_0^4 - 2x_0^2 y_0^2 + 4x_0^2 y_0^2} \\
&= \sqrt{(x_0^2 + y_0^2)^2} \\
&= |z_0|^2
\end{aligned}
$$

Hence $|F(z_0)| < |z_0|$ as long as $0 < |z_0| < 1$. It follows that one iteration of F moves z_0 closer to 0. Continuing this process (and using the fact that repeated squaring of a number $|z_0| < 1$ tends to 0) we see that any z_0 with $|z_0| < 1$ has orbit which tends to 0. Thus all of these orbits are stable.

Similar arguments show that, if $|z_0| > 1$, then the orbit of z_0 tends to ∞ under iteration. Hence all points z_0 in the plane with $|z_0| \neq 1$ have stable orbits. What about points with $|z_0| = 1$? These are points which lie on the unit circle in the plane. Clearly, any such point lies in the chaotic set, since points just outside the circle have orbits tending to ∞ while points just inside have orbits tending to 0.

The reader may object that this does not seem to be too chaotic a situation. After all, we know exactly what is happening. All points off the unit circle have predictable behavior, and all points on the unit circle have orbits which remain there forever. Indeed, if $|z_0| = 1$, then

$$|F(z_0)| = (x_0^2 + y_0^2)^2$$

as we saw above, so $|F(z_0)| = 1$ as well. However, it is an amazing fact that the dynamics of F on the unit circle are *exactly* the same as those of $F(x) = 4x(1-x)$ on the interval $0 \leq x \leq 1$! While we will not prove this here, the proof is quite simple and we refer the reader to [25] for more information about this.

Thus we see that the complex plane decomposes into two regions: the stable set consisting of those points with orbits tending to 0 and ∞, and the chaotic set which is the unit circle. This is typical, as we shall see, for complex maps. The chaotic set, the object of our interest, is called the *Julia set*, after the French mathematician Gaston Julia who first studied this set in the early twentieth century.

3.3.2 The Julia set

How do we use computer graphics to find the Julia set? Actually, there are several ways. The first, and easiest to program and to compute, is the *inverse iteration method* (see IIM and its variations in Chapter 4). This method works as follows. Let us work with $F(z) = z^2$ for the moment. Given z_0, we ask which points map to z_0 under F. Naturally, these points are the square roots of z_0, and, just as in the real case, there are two of them. To compute the two square roots of a complex number z_0, we need to introduce the *polar representation* of z_0. In this representation, z_0 is determined by the polar angle and distance from z_0 to the origin. The polar angle is the angle that the ray from the origin to z_0 makes with the positive x-axis measured in a counterclockwise direction (usually in radians). Negative angles are measured in the clockwise direction.

We let θ_0 denote the polar angle of z_0. See Figure 3.5. The radius of z_0 is simply its absolute value $r_0 = |z_0|$. The two numbers r_0 and θ_0 completely specify z_0.

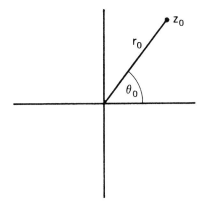

Fig. 3.5: Polar coordinates of the complex number Z_0

Now how do we compute $\sqrt{z_0}$? The two square roots are given in terms of the polar representation (r_0, θ_0) of z_0. The new polar radius is simply $\sqrt{r_0}$ (where we choose the positive square root) and the two new angles are $\theta_0/2$ and $\pi + \theta_0/2$.

For example, if $z_0 = i$, then $r_0 = 1$, $\theta_0 = \pi/2$. Thus $\sqrt{i}$ has the polar representation either $r_0 = 1$, $\theta_0 = \frac{\pi}{4}$ or $r_0 = 1$, $\theta_0 = \frac{\pi}{4} + \pi$.

To compute the backward orbit of z_0, we simply take successive square roots. At each stage we are free to choose one of two polar angles. Hence there are lots of different possible backward orbits of z_0 (as long as $z_0 \neq 0$). The amazing fact is that any of them converges to the Julia set of F! That is if we plot successive square roots of $z_0 \neq 0$, omitting the first few points along the backward orbit. we produce the set of unstable orbits which, in this case, is just the unit circle[1].

The Julia set for any function of the form $F(z) = z^2 + c$ may be obtained in precisely the same way. The displayed algorithm *JuliaIIM()* gives a simple BASIC program to plot the Julia set of $z^2 + c$ for various c-values. Note that this can be interpreted as an example of an iterated function system (IFS) with $\{w_n, p_n : n = 1, 2\}$ where

$$w_1 = +\sqrt{z - c} \quad \text{and} \quad w_2 = -\sqrt{z - c}$$

and $p_1 = p_2 = 0.5$ (see Chapter 5). Here we note that c may be complex and that the window in the plane is given by $|x| \leq 2$, $|y| \leq 2$. Note the fascinatingly

---

[1] We remark that the dynamics of $z \mapsto z^2$ on the unit circle is chaotic, i.e. there are dense orbits, repelling periodic points densely fill the circle and the dynamics is mixing.

```
ALGORITHM JuliaIIM ()
Title            BASIC program to compute Julia sets by inverse iterations

INPUT "cx, cy", cx, cy
INPUT "x, y", x, y
FOR i=1 TO 6000
     wx = x - cx
     wy = y - cy
     IF wx > 0 THEN theta = ATN (wy / wx)
     IF wx < 0 THEN theta = 3.14159 + ATN (wy / wx)
     IF wx = 0 THEN theta = 1.57079
     theta = theta / 2
     r = SQR (wx * wx + wy * wy)
     IF RND < 0.5 THEN r = SQR (r) ELSE r = - SQR (r)
     x = r * COS (theta)
     y = r * SIN (theta)
     m = -5 + (x + 4) * 500 / 8
     n = (2 - y) * 250 / 4
     CALL MOVETO (m, n)
     CALL LINE (0, 0)
NEXT i
END
```

different shapes these Julia sets assume! The reader should compare the Julia sets for $c = -1, c = -2, c = 0, c = 0.2 + 0.3i, c = i$, some of which are depicted in Figure 3.6 below.

3.3.3   Julia sets as basin boundaries

There is another method that is commonly used to plot Julia sets of polynomials like $F(z) = z^2 + c$ (see the Boundary Scanning Method BSM and its variants in Chapter 4). This method plots all points which under iteration of F do not escape to ∞. It is a fact that the boundary or frontier of this set of non-escaping points is precisely the Julia set.

We have plotted several of these regions in Figure 3.7 . These plots were obtained as follows: each point in a 1024×1024 grid defined by $|x|, |y| < 2$ was iterated up to 30 times. If $|x|$, $|y| < 2$ and $|F^n(x + iy)| > 3$ for some n, then it can be proved that such a point has an orbit which tends to ∞. Such a point therefore does not lie in the Julia set. We have colored such points white. On the other hand, points whose first 30 iterates remain within the circle of radius 3 are colored black. It can be proved that points whose orbits remain inside this circle for *all* iterations lie in the so-called filled-in Julia set , i.e., in the set whose boundary is precisely the Julia set.

As a remark, we note that the filled-in Julia set takes much, much longer

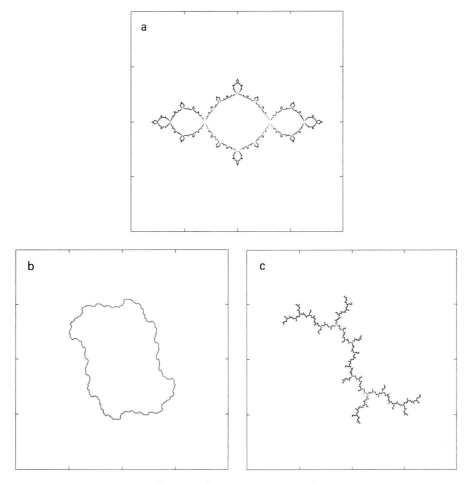

Fig. 3.6: The Julia sets for **a.** $z^2 - 1$, **b.** $z^2 + 0.2 + 0.3i$ and **c.** $z^2 + i$. The pictures show 100,000 points of backward orbits in the region $-2 \le \operatorname{Re} z \le 2$ and $-2 \le \operatorname{Im} z \le 2$

to compute than the simple Julia set of $z^2 + c$ as described in the last section. Somehow, though, these pictures are more appealing! Perhaps it is the length of time it takes to compute, or perhaps the proportion of black to white which results on the screen, but, for whatever reason, filled-in Julia sets make much more interesting pictures. However, mathematically speaking, the previous algorithm gives precisely the set of points which comprise the chaotic region.

Figure 3.7 shows the result of applying this algorithm to depict the Julia sets of various quadratic functions.

3.3.4 Other Julia sets

The next two sections of this paper are significantly more mathematical than the previous sections. They represent some of the recent advances in the field

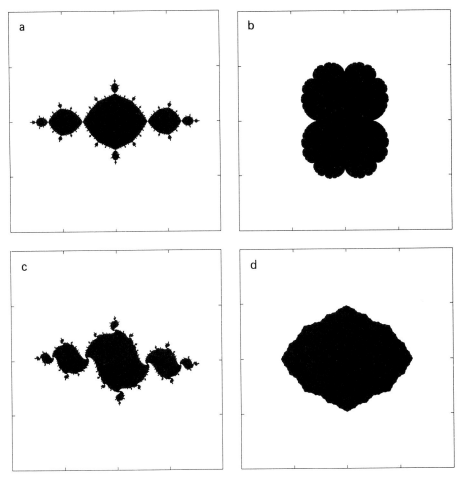

Fig. 3.7: The Julia sets of **a**. $z^2 - 1$, **b**. $z^2 + .25$, **c**. $z^2 - 0.9 + 0.12i$, and **d**. $z^2 - 0.3$. The region shown is $-2 \leq \text{Re } z \leq 2$, $-2 \leq \text{Im } z \leq 2$

of complex analytic dynamics. The reader who is not mathematically inclined may choose to skip many of the technical details; however, the algorithms that generate the Julia sets of these maps are just as accessible as those in previous sections.

Any complex analytic function has a Julia set. As with the quadratic family, the Julia sets of other complex functions are often fractals. Besides polynomials, the classes of maps most often studied are rational maps (i.e., quotients of two polynomials) and entire transcendental functions (i.e., functions with no singularities in the finite plane and which are not polynomials). In this section we will describe a method for computing the Julia sets of this latter class.

First, everything we say will be true for the class of entire transcendental functions which are of *finite type*, that is, for maps with only finitely many critical values and asymptotic values. A point q is a *critical value* if $q = F(p)$ where

$F'(p) = 0$. A point q is an *asymptotic value* if there is a curve $\alpha(t)$ satisfying

$$\lim_{t \to 1} \alpha(t) = \infty \quad \text{and} \quad \lim_{t \to 1} F(\alpha(t)) = q.$$

A typical example of an asymptotic value is the omitted value 0 for the complex exponential function.

Since e^z has no critical values and only one asymptotic value, this map is of finite type. Also, the maps

$$\begin{aligned} \sin z &= \frac{1}{2i}(e^{iz} - e^{-iz}) \\ \cos z &= \frac{1}{2}(e^{iz} + e^{-iz}) \end{aligned}$$

have infinitely many critical points but only two critical values, the maximum and minimum values of $\sin x$ and $\cos x$ as functions on the real line.

In this section we will deal only with λe^z, $\lambda \sin z$, and $\lambda \cos z$. The main reason for this is that these examples are typical for maps of finite type and the computer graphics associated with these maps are quite spectacular.

One reason for the importance of these maps is that the Julia set has an alternative characterization which makes it quite easy to compute. For maps like λe^z, $\lambda \sin z$, and $\lambda \cos z$, the Julia set is also the closure of the set of points whose orbits tend to infinity. Note that this is quite different from the definition of the Julia set for polynomials (see Chapter 4). In the latter case, the Julia set formed the boundary of the set of escaping orbits. In the entire case, points whose orbits escape actually lie in the Julia set.

The reader may protest that there is nothing apparently chaotic about an orbit tending to ∞. However, in the entire case, ∞ is an essential singularity, which is the main factor distinguishing these maps from polynomials. By the Great Picard Theorem, any small neighborhood of ∞ is smeared infinitely often over the entire plane and this suggests that orbits tending to ∞ experience quite a bit of stretching. Also, the Julia set is the *closure* of the set of escaping orbits. That is, the Julia set contains not only the escaping points, but also any limit point of a sequence of such points. Included among these limit points are all of the repelling[2] periodic points, so the Julia set is by no means solely comprised of escaping points.

Thus to produce computer graphics images of Julia sets of entire functions like e^z or $\sin z$, we need only look for points whose orbits go far from the origin. Generally, orbits tend to ∞ in specific directions. For example, for λe^z, orbits

---

[2] For a definition see Section 4.2 .

which escape tend to do so with increasing real part. That is, if $E_\lambda(z) = \lambda e^z$ and

$$\lim_{n \to \infty} |E_\lambda^n(z)| = \infty$$

then

$$\lim_{n \to \infty} \text{Re}\,(E_\lambda^n(z)) = \infty$$

For $S_\lambda(z) = \lambda \sin z$, the situation is quite different. If

$$\lim_{n \to \infty} S_\lambda^n(z) = \infty$$

then

$$\lim_{n \to \infty} |\text{Im}\,(S_\lambda^n(z))| = \infty.$$

So orbits tend to ∞ for E_λ in the direction of the positive real axis, while for S_λ they tend to ∞ along the positive or negative imaginary axis.

These facts allow us to use a simple test to compute the Julia sets of these maps. For E_λ, we simply test whether an orbit ever contains a point with real part greater than 50 (since e^{50} is quite large). A similar test works using the absolute value of the imaginary part for $\sin z$. All of the pictures of the Julia sets of λe^z and $\lambda \sin z$ were generated using this algorithm, with the exception of Fig. 3.8, which uses a modification.

To be more specific, let us outline the algorithm that produces the Julia set of $\sin z$, as depicted in Figure 3.10. If we write $z = x + iy$ and $\sin z = x_1 + iy_1$, then

$$\begin{aligned} x_1 &= \sin x \cosh y \\ y_1 &= \cos x \sinh y. \end{aligned}$$

This gives us the formula to compute the main body of the loop in the program. If $|y_1| > 50$ at any stage in the , then we "pretend" that the original point has escaped to ∞ and color it white. On the other hand, if the original point never escapes, we color it black. These points do not lie in the Julia set. We used 25 iterations to draw the Julia set in Figures 3.10 .

The computer graphics images of transcendental functions like $\sin z$ and $\exp z$ are quite attractive when color is used. In this case, it is best to color the complement of the Julia set black and to color the Julia set itself with colors which depend upon the rate of escape to ∞. In the graphics presented in the color plates, red was used to denote points which escape to ∞ fastest. Shades of orange, yellow and green were used to color points which escaped less quickly. And shades of blue and violet were used for points which escaped, but only after a significant number of iterations. These images were computed at a resolution of 1000×800.

3.3.5 Exploding Julia sets

The computer graphics which accompanies the study of entire functions reveals pictures of incredible beauty as well as some very startling mathematics. For example, computer graphics reveals that the Julia sets of these maps may literally explode. Consider the family $E_\lambda(z) = \lambda e^z$ where $\lambda \in \mathbf{R}$. Figure 3.8 depicts the Julia set of $\frac{1}{e}e^z$. Note the large black region. As in our previous examples black points correspond to points with stable orbits. In this case, all black points may be shown to have orbits which tend to a fixed point at 1. On the other hand, in Figure 3.9 we have plotted the Julia set of λe^z when $\lambda = \frac{1}{e} + 0.1$. Note the dramatic change in the character of the black region! In fact, one may prove that the Julia set of E_λ when $\lambda > \frac{1}{e}$ is the entire complex plane, while, when $0 < \lambda \le \frac{1}{e}$, the Julia set omits the entire half plane Re $z < 1$.

A similar phenomenon occurs in the sine family $S_\lambda(z) = \lambda \sin z$ when $\lambda = 1 + \varepsilon i$. Figure 3.10a displays the Julia set of $\sin z$, i.e., when $\varepsilon = 0$. Figure 3.10b and 3.10c display the case when $\varepsilon > 0$. One may prove that there is a curve of parameter values which approaches $\varepsilon = 0$ in the direction of the imaginary axis on which the Julia set of S_λ is the entire plane. That this curve is the straight line $\lambda = 1 + \varepsilon i$ is unlikely, but one does at least get a sense of the explosion in this case.

The Julia sets of both E_λ and S_λ are fractals, at least before the explosions. However, these fractal sets have Hausdorff dimension 2. Nevertheless, they do not contain open sets but rather are intricate patterns called Cantor bouquets (roughly speaking, the cross product of a Cantor set and a curve). Note the apparent dissimilarity between the Cantor bouquets for E_λ in Figure 3.8 and that of S_λ in Figure 3.10. The Julia set of E_λ does resemble a collection of curves whereas the Julia set for S_λ does not. The reason for this is a rather remarkable recent result of Curt McMullen which asserts that the Lebesgue measure of the Julia set of E_λ is zero while for S_λ it is positive and quite large. For further reading on these experiments, we refer the reader to [26].

Let us try to explain why these dramatic changes in the structure of the Julia sets occur. As we discussed in the previous section, the Julia sets of these maps are the closure of the set of points whose orbits escape to infinity. It is the sudden occurrence of the escape of all singular values of the map which causes the explosion.

To be more precise, a *singular value* is either a critical value or an asymptotic value. For λe^z, there is only one asymptotic value, the omitted value 0, while $\lambda \sin z$ and $\lambda \cos z$ have two critical values and no asymptotic values. The graph

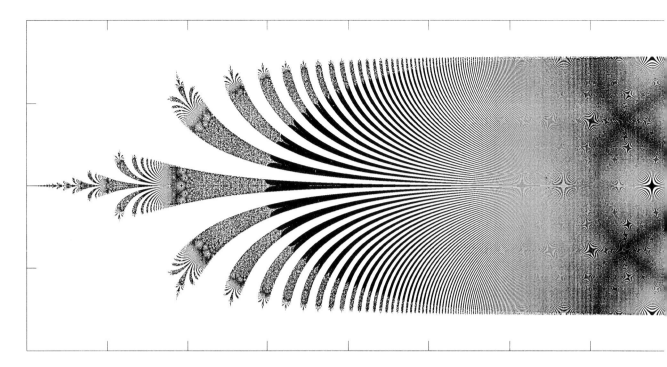

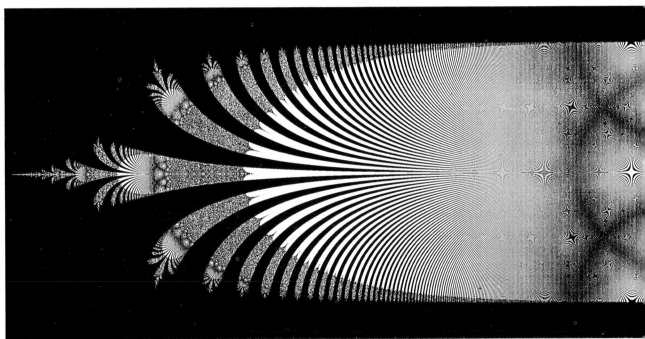

Fig. 3.8: In the bottom figure the Julia set of $\frac{1}{e}e^z$ (in white) approximated by points that escape
to ∞ according to the following criteria: A point z_0 is colored white if the orbit of z_0 contains a
point z_k with real part Re z_k greater than 50 within the first 25 iterates and if $\cos(\operatorname{Im} z_k) \geq 0.0$
(i.e. the next iterate z_{k+1} will be very large and is *not* in the left half plane Re $z \leq 0$ which
can easily be shown not to be part of Julia set). The image shows the region $1 \leq$ Re $z \leq 9$,
$-2 \leq \operatorname{Im} z \leq 2$. The strange artifacts at the right side of the figures are due to aliasing defects,
also called Moiré patterns . The top figure shows the image in reversed colors.

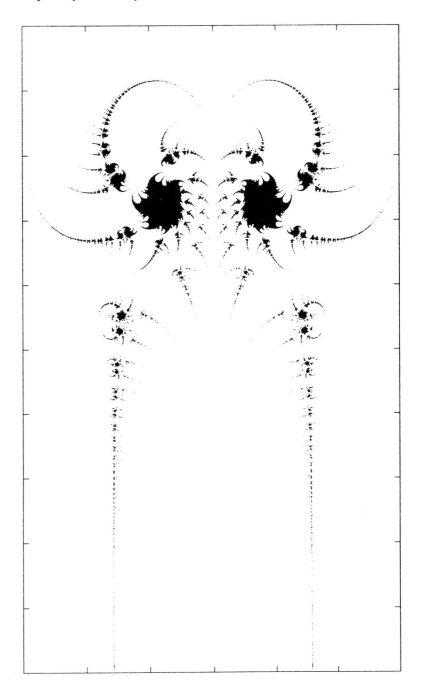

Fig. 3.9: The Julia set of $(\frac{1}{e} + 0.1)\,e^z$ (in white). The displayed region is $-2 \leq \operatorname{Re} z \leq 8$ and $-3 \leq \operatorname{Im} z \leq 3$. The image is rotated by $90°$ clockwise.

of λe^z is displayed in the two cases $\lambda < \frac{1}{e}$ and $\lambda > \frac{1}{e}$ in Figure 3.11 . Note that when $\lambda < \frac{1}{e}$, the orbit of 0 does not escape to ∞ but rather is attracted by the attracting fixed point at q. (This is a general fact in complex dynamics: if there is an attracting periodic orbit for the map, then this orbit must attract at least one

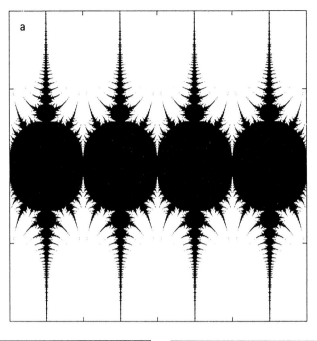

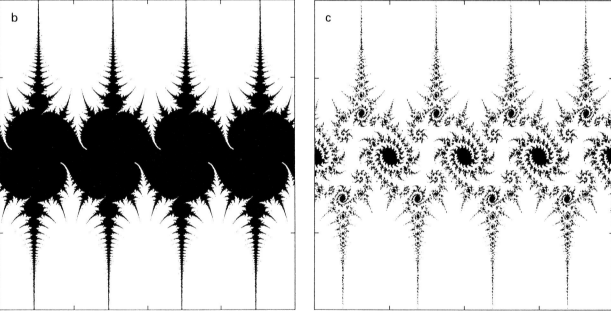

Fig. 3.10: The Julia set of $S_\lambda(z) = \lambda \sin z$ where $\lambda = 1 + \varepsilon i$ in is shown in white. The region is $-2\pi \leq \text{Re } z \leq 2\pi$, $-2\pi \leq \text{Im } z \leq 2\pi$. **a.** $\varepsilon = 0$, **b.** $\varepsilon = 0.1i$, **c.** $\varepsilon = 0.4i$

of the singular values[3].) When $\lambda > \frac{1}{e}$, the two fixed points for λe^z disappear from the real line and $E_\lambda^n(0) \to \infty$.

---

[3] See e.g. the discussion in Section 4.2.2 .

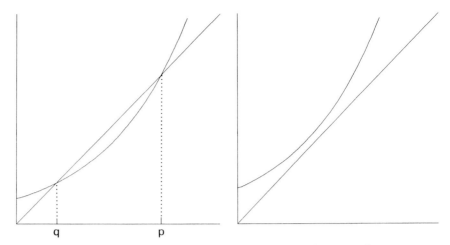

Fig. 3.11: The graphs of $S_\lambda(z) = \lambda e^z$ where $\lambda < \frac{1}{e}$ and $\lambda > \frac{1}{e}$.

The fact that the escape of all singular values implies that the Julia set explodes is a recent theorem of D. Sullivan, and was extended to the entire case by L. Goldberg and L. Keen. What Sullivan's theorem actually says is, if there is any component of the complement of the Julia set, then it must have associated with it at least one singular value. Thus, if all singular values tend to ∞ (and thus are in the Julia set), there can be no stable behavior whatsoever.

A word about the bifurcation which occurs for λe^z at $\lambda = \frac{1}{e}$. At this value the two fixed points q and p coalesce into one neutral fixed point. This is called the *saddle node bifurcation* . When $\lambda > \frac{1}{e}$, it seems that these fixed points disappear. However, these fixed points have become complex. Indeed, they form the central points of the main spiral in Figure 3.9. The reason that we see black blobs in this picture is that these fixed points are very weakly repelling – it takes a long time for nearby orbits to go far away. Since our algorithm only computed the first 25 points along an orbit, certain nearby orbits had not yet escaped. If one were to run the same program with 50 or 100 or 200 iterations, one would see a gradual shrinking of the black region. On the other hand, no larger amount of iterations would change Figs. 3.8 or 3.10 .

There are other examples of similar explosions in the dynamics of complex functions. For example, the function $i\lambda \cos z$ experiences a similar saddle-node explosion as λ increases through 0.67.

3.3.6 Intermittency

There are many, many fascinating phenomena that occur in the dynamics of entire functions. Here we present another typical event, intermittency. Recall that in the last section we asked what happens if all singular values for the map escaped. The answer was an explosion. Here we ask the related question what happens if singular values go far away but then return.

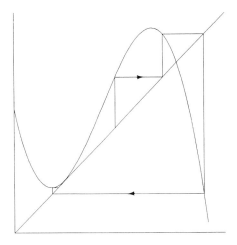

Fig. 3.12: The graph of P_0.

The typical situation occurs when we have a graph as in Figure 3.12 . Here the function P_0 has a saddle-node fixed point at q. Note that the orbit of p depicted at first escapes, but then returns to be attracted by q. This is an example of a *homoclinic orbit*. If we perturb this map to

$$P_{-\varepsilon}(x) = P_0(x) - \varepsilon$$

then we clearly produce two new fixed points, one attracting and one repelling. On the other hand, for $P_\varepsilon(x) = P_0(x) + \varepsilon$ there are no fixed points on the real line. As in the last section, as ε increases through 0, the family P_ε undergoes a saddle-node bifurcation. See Figure 3.13 .

There is a difference between this bifurcation and the saddle-node bifurcation for λe^z described in the previous section. In this case, the critical point whose orbit tended to the attracting fixed point when $\varepsilon \leq 0$, now escapes far away but then returns. Associated with this phenomenon is a wealth of dynamical behavior that is only beginning to be understood.

Let us now try to choose various $\varepsilon > 0$ values to illustrate the different types of orbits that may occur. Figure 3.14a shows that, for certain ε-values, the critical point itself may be periodic. If this is the case, it is known that this orbit must be attracting. In the computer pictures, such an ε-value would lead to a cycle of black regions mapped one to the other. Figure 3.14b shows that we will be able to find ε-values much smaller so that the critical value is periodic with much higher period. But there must be other kinds of behavior as well. For

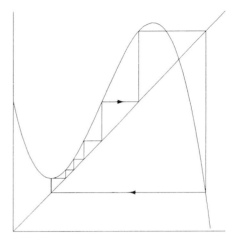

Fig. 3.13: The graph of P_ε.

example, Figure 3.14c shows that the critical value may return nearby but not land on itself. In this case, the critical orbit could cycle around aperiodically. Finally, Figure 3.14d shows that the critical point could actually go far away and never return.

This phenomenon is called intermittency. The key to understanding it lies in the Mandelbrot set, for it is known that in any interval of the form $(0, \varepsilon_0)$, the parameter ε passes through infinitely many Mandelbrot-like sets. In each passage, the dynamics of P_ε undergoes (locally) all of the changes associated with the quadratic dynamics described in Sections 3.3.2-3.3.3 (and Chapter 4).

The color plates associated with this describe this event for the family $C_\lambda(z) = \lambda \cos z$ when λ decreases from π to 2.92. The graphs of C_λ are depicted in Figure 3.15 . Note the similarity with Figs. 3.12 and 3.13 . In the color figures we recall that black regions represent points not in the Julia set whereas colored points represent points whose orbit escapes. Red points are those whose orbits escape most quickly, followed in order by shades of orange, yellow, green, blue, and violet. A total of 35 iterations were used to create these images, so that an orbit whose imaginary part remained between ± 50 for all 35 iterations was colored black. Each image represents a 1000 by 800 grid in the complex plane centered at $-\pi$ (the sides of these rectangles vary). Note that $-\pi$ is the critical value and the attracting fixed point for C_π. The large oval black region in Plate 17 is the basin of attraction for $-\pi$. The satellite black regions represent preimages of this basin.

As λ decreases to approximately 2.97, there remains an attracting fixed point for C_λ. At 2.97 the saddle-node occurs;the basin for this neutral fixed

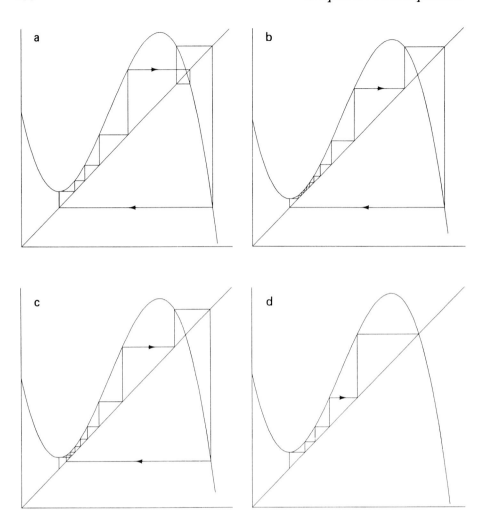

Fig. 3.14:

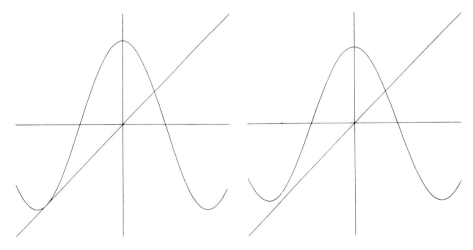

Fig. 3.15: The graphs of C_λ.

point is depicted in Plate 18 . Note the similarity between this basin and that of $z \mapsto z^2 + 0.25$ depicted in Figure 3.7 .

As λ decreases through 2.97, the Julia set changes dramatically. See Plates 19 and 20 . Note the presence of black regions whose locations have changed slightly as the parameter decreases. Actually, there is much more happening in these black regions than what appears in these plates. The remaining four plates show enlargements of some of these regions for smaller parameter values. Note that the original black bubbles actually have structures within that are reminiscent of the original bifurcations (Plates 19 and 20) . This, of course, is characteristic of fractal behavior.

This entire sequence of bifurcations has been put on videotape by computing a continuous selection of λ-values between π and 2.92. The resulting film, "Exploding Bubbles," illustrates how these black regions move and change as the parameter is varied. Copies of this and other films of complex bifurcations are available from the author.

Chapter 4

Fantastic deterministic fractals

Heinz-Otto Peitgen

4.1 Introduction

The goal of this chapter, which is a continuation of Section 3.3, is to demonstrate how genuine mathematical research experiments open a door to a seemingly inexhaustible new reservoir of fantastic shapes and images. Their aesthetic appeal stems from structures which at the same time are beyond imagination and yet look extremely realistic. Being the result of well defined and mostly very simple mathematical processes — which depend on a few parameters — animations of evolution or metamorphosis of fantastic forms are easy to obtain.

The experiments which we present here are part of a world wide interest in *complex dynamical systems*. They deal with chaos and order and with their competition and coexistence. They show the transition from one to another and how magnificently complex the transitional region generally is.

The processes chosen here come from various physical or mathematical problems. They all have in common the competition of several centers for the domination of the plane. A single boundary between territories is seldom the result of this contest. Usually, an unending filigreed entanglement and unceasing bargaining for even the smallest areas results.

Our pictures represent processes, which are, of course, simplified idealizations of reality. They exaggerate certain aspects to make them clearer. For example, no real structure can be magnified repeatedly an infinite number of times and still look the same. The principle of self-similarity (see 1.1.2 and 1.1.4) is nonetheless realized approximately in nature: In coastlines and riverbeds, in cloud formations and snow flakes, in turbulent flow of liquids and the hierarchical organization of living systems. It was Benoit B. Mandelbrot who has opened our eyes to the "fractal geometry of nature".

In the first paragraph we discuss the most important model example, i.e. the feedback process given by $z \mapsto z^2 + c$ in the complex plane. Particular attention is given to a background and graphics discussion of

- Julia sets, and the

- Mandelbrot set.

We believe that some understanding of the mathematical background will allow the reader to see some of the most popular algorithms like "Mandelzoom" in [27,28] in a new perspective which easily provides a basis to go on and beyond and develop own ideas and extensions/improvements. Furthermore, this approach seems to be the most effective way to discuss and present various algorithms. Unfortunately, a mathematical presentation of an algorithm leaves a lot of details to the actual implementation and requires some good familiarity with its parameters. Where appropriate, we will give some hints towards a concrete implementation, usually in footnotes. The second paragraph will briefly discuss generalizations and extensions such as

- rational mappings of the complex plane;

- mappings of real Euclidean space, in particular, Newton's map.

The third paragraph collects some information on particular problems from the point of view of graphical realizations, e.g. questions of animation.

4.2   The quadratic family

This paragraph is entirely devoted to a discussion of various fractal aspects involved in the mapping

$$f_c : \mathbf{C} \to \mathbf{C} , f_c(z) = z^2 + c, \ c \in \mathbf{C} .$$

($\mathbf{C}$ denotes the complex number field.) The dynamics of f_c is an enormously rich fountain of fractal structures. Sometimes it will be useful to think of the complex plane being projected onto a 2-sphere S^2, so that each point Z on the sphere corresponds to one and only one point $\pi(Z)$ in the plane, except for the north pole, which we interpret as the point "at infinity".

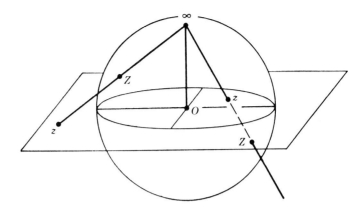

Fig. 4.1: Stereographic projection

In this sense we look at S^2 as $\mathbf{C} \cup \{\infty\}$. The action of f_c on $\mathbf{C}$ translates to an action on S^2, i.e. orbits

$$z \in \mathbf{C} : z \mapsto f_c(z) \mapsto f_c(f_c(z)) \mapsto \dots$$

correspond to

$$Z \in S^2 : Z \mapsto \tilde{f}_c(Z) \mapsto \tilde{f}_c(\tilde{f}_c(Z)) \mapsto \dots$$

orbits on S^2, where

$$\tilde{f}_c(Z) = \pi^{-1}(f_c(\pi(Z))).$$

To discuss the behavior of f_c near ∞ one usually invokes another transformation, the reflection $r(z) = \frac{1}{z}$ at the unit circle, which exchanges 0 with ∞; i.e. one looks at

$$F_c(z) = r(f_c(r(z))) = r \circ f_c \circ r(z) \qquad (4.1)$$

near 0 in order to understand f_c near ∞. One calculates that

$$F_c(z) = \frac{z^2}{1 + cz^2}. \qquad (4.2)$$

Now $F_c(0) = 0$ for all $c \in \mathbf{C}$ and, moreover, $F_c'(0) = 0$. That means that 0 is an *attracting*[1] fixed point for F_c, or, equivalently ∞ is an attracting fixed point

---

[1]The norm of a complex number z is defined as $|z| = \sqrt{\text{Re}(z)^2 + \text{Im}(z)^2}$, where $\text{Re}(z)$ and $\text{Im}(z)$ denote the real and imaginary parts of z. Let R be a mapping on $\mathbf{C}$ and $R(z) = z$. Then z is a *fixed point* and

- z is *attracting*, provided $|R'(z)| < 1$
- z is *repelling*, provided $|R'(z)| > 1$
- z is *indifferent*, provided $R'(z) = \exp(2\pi i \alpha)$.

In the latter case α may be *rational* or *irrational* and accordingly z is called rationally or irrationally indifferent. If $R^k(z) = R(R^{k-1}(z)) = z$ and $R^j(z) \neq z$ for $1 \leq j < k$ we call z a *point of period k* and the set $\{z, R(z), \dots, R^{k-1}(z)\}$ is called a *k-cycle*, which may be attracting, repelling or indifferent, depending on $|(R^k)'(z)|$.

for f_c. Naturally, attracting means that points z_0 near ∞ will generate orbits

$$z_0 \mapsto z_1 \mapsto z_2 \mapsto z_3 \ldots$$

$z_{k+1} = f_c(z_k)$, $k = 0, 1, \ldots$, which approach ∞. Collecting all such points one obtains the *basin of attraction* of ∞:

$$A_c(\infty) = \{z_0 \in \mathbf{C} : f_c^k(z_0) \to \infty \text{ as } k \to \infty\}, \qquad (4.3)$$

where $f_c^k(z_0) = f_c(f_c^{k-1}(z_0)) = z_k$. Naturally, $A_c(\infty)$ depends on c. Figure 4.2 illustrates the situation.

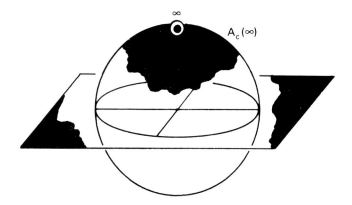

Fig. 4.2: Basin of attraction of ∞ in black.

(Readers who feel uncomfortable with ∞ may of course work with F_c and discuss the point 0.) $A_c(\infty)$ has a natural boundary, i.e. there are always points z_0, which generate orbits which do not approach ∞, thus stay bounded. This is seen immediately, because, for example, f_c has always two fixed points (i.e. solutions of $z^2 + c = z$). The boundary of $A_c(\infty)$ is denoted by $\partial A_c(\infty)$ and is called the *Julia*[2] *set* of f_c. We also use the symbol $J_c = \partial A_c(\infty)$. This is the central object of interest in the theory of iterations on the complex plane. Together with $A_c(\infty)$ and J_c goes a third object (sometimes called the *filled-in Julia set*) :

$$K_c = \mathbf{C} \setminus A_c(\infty) = \left\{ z_0 \in \mathbf{C} : f_c^k(z_0) \text{ stays bounded for all k} \right\} \qquad (4.4)$$

Obviously, we have that

$$\partial K_c = J_c = \partial A_c(\infty), \qquad (4.5)$$

---

[2] Gaston Julia (1893 - 1978) was a French mathematician, who together with Pierre Fatou (1878-1929) laid the basis for the beautiful theory discussing iterations of rational mappings in the complex plane.

i.e. J_c separates competition between orbits being attracted to ∞ and orbits remaining bounded as $k \rightarrow \infty$. There are two cases of c for which J_c is very simple:

$$c = 0 \quad : \quad J_c = \{z : |z| = 1\} \quad \text{"unit circle"}$$
$$c = -2 \quad : \quad J_c = \{z : -2 \leq Re(z) \leq +2, Im(z) = 0\} \quad \text{"interval"}$$

(While the case for $c = 0$ is immediate, the case for $c = -2$ is not so obvious; see [83]). For all other values of c the Julia set J_c is an amazingly complicated set, a fractal indeed. Moreover, the J_c's change dramatically as c changes. Figure 4.3 shows a collection of 8 different J_c's for 8 choices of c. In general, it is not obvious at all how to obtain a reasonable picture of J_c, though there is an immediate algorithm (IIM - Inverse Iteration Method) obtained from the following characterization due to Julia and Fatou: For any c the equation $f_c(z) = z$ has two finite solutions $u_0 \neq \infty \neq v_0$ - fixed points. If $c \neq \frac{1}{4}$ then one of them is a repelling[3] fixed point, say u_0.

Then one has

$$J_c = \text{closure}\left\{z : f_c^k(z) = u_0 \text{ for some integer } k\right\}. \tag{4.6}$$

Note that in general $f_c^k(z) = u_0$ has 2^k solutions, i.e. the total number of iterated preimages of u_0 obtained by recursively solving the equation $z^2 + c = a$ is

$$n(k) = 2^{k+1} - 1, \quad k = 0, 1, \dots$$

(beginning with $a = u_0$ yielding u_1 and u_2, then $a = u_1, a = u_2$ yielding u_3, u_4, u_5, u_6, etc., see Chapter 3 for solving equations in C). The recursion is nicely represented in a tree.

Figure 4.4a shows a labeling of the $n(k)$ knots, representing solutions in the recursion, which is induced naturally. The notation "closure" in Eqn. (4.6) means that each point in J_c can be obtained as a limit of a sequence composed by points from the iterated preimages. Let

$$J_{c,k} = \left\{u_0, u_1, \dots, u_{n(k)-1}\right\}$$

be the collection of all iterated preimages up to level k beginning with u_0. Then another way to express Eqn. (4.6) is to say that for k sufficently large $J_{c,k}$ and J_c will be indistinguishable, i.e. the iterated preimages are dense in J_c. In fact a more general statement is true: If z_0 is just any point in J_c, then

$$J_c = \text{closure}\left\{z : f_c^k(z) = z_0 \text{ for some integer } k\right\}. \tag{4.7}$$

---

[3] If $c = 1/4$ then both fixed points are neither repelling nor attracting. In fact (see section on Sullivan classification) in this case $u_0 = v_0$ and both are parabolic fixed points i.e. $f_c'(u_0) = e^{2\pi i \alpha}$ with α rational (here $\alpha = 0$), and it is known that parabolic fixed points are in the Julia set.

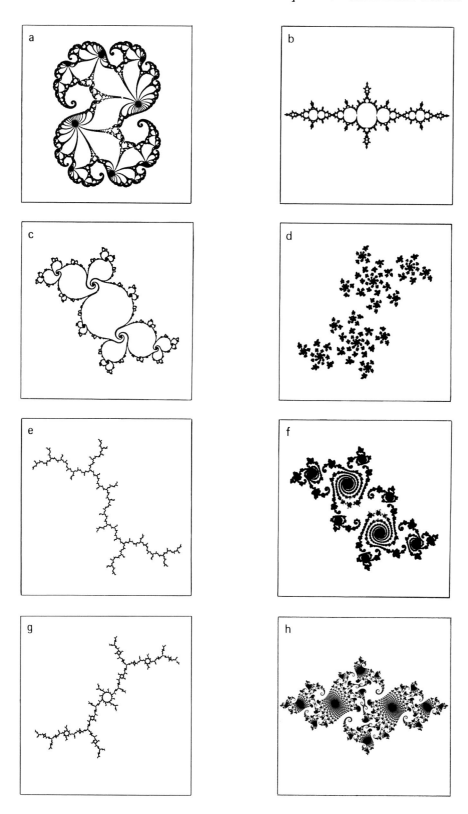

Fig. 4.3: 8 Julia sets

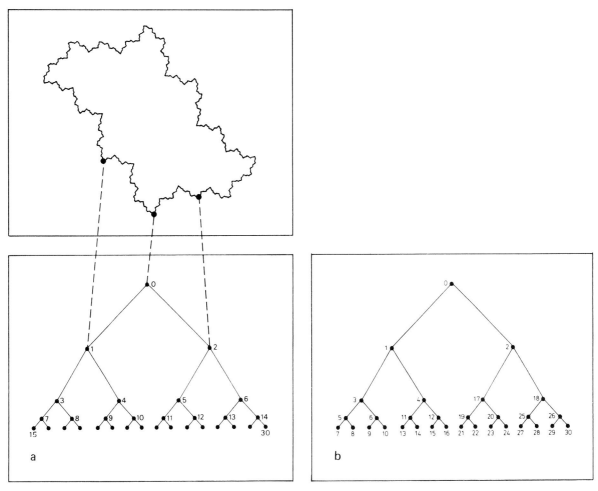

Fig. 4.4: Tree-structure in recursion

Also, there is the characterization

$$J_c = \text{closure } \{\text{all repelling periodic points of } f_c\}, \tag{4.8}$$

which provides, for example, many choices for a "seed" z_0 in J_c.

The IIM - Algorithms take advantage of (4.6) or (4.7) and work nicely for a number of J_c's. Though $J_{c,k}$ will approximate J_c there are cases where the iterated preimages fill J_c "infinitly slowly". The reason is that the iterated preimages in many cases do not uniformly cover J_c, i.e. "hit" certain parts of J_c very frequently while other parts of J_c are "hit" extremely rarely. Figure 4.5 shows a Julia set and the histogram of the density distribution (vertical white bars) of IIM in the upper right. Here are some of difficult c-values :

$$c = +0.27334 + 0.00342\,i,$$
$$c = -0.48176 - 0.53165\,i.$$

This is why variations of IIM (see MIIM) or totally different methods — like
the BSM (=Boundary Scanning Method) — are necessary.

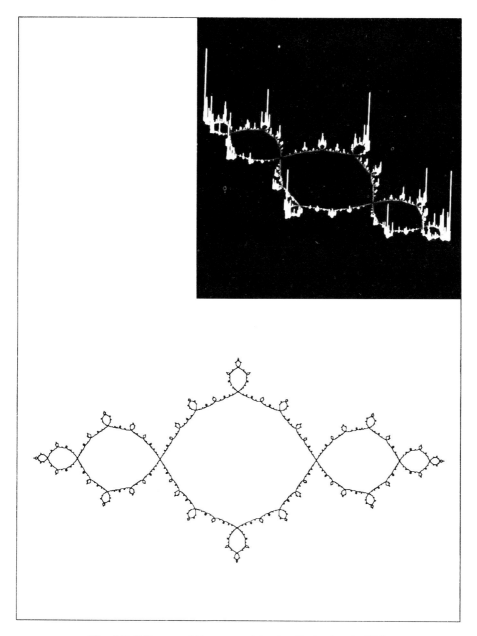

Fig. 4.5: Julia set and histogram of density distribution for IIM

The BSM-Algorithm is even more elementary than IIM. It uses the definition
of K_c, Eqn. (4.4) , and $A_c(\infty)$, Eqn. (4.3) , in a straightforward manner.

The IIM- and BSM-Algorithms are the basic algorithms for Julia sets, which
easily generalize to other dynamical systems.

IIM - Inverse Iteration Methods : Suppose we have $u_0 \in J_c$, for example if u_0 is the repelling fixed point of f_c.

(A) The most simple algorithm (see Chapter 5) is obtained by choosing one of the two roots at random at each stage of the recursion for preimages. This amounts to a random walk on the tree in Figure 4.4a. Usually the method will work for almost all initial u_0 in C (e.g. for $c = 0$ one obviously has to avoid $u_0 = 0$). The first few preimages will have to be excluded from the plot. Iterated preimages again will approximate J_c. Formally, this is a typical example for an IFS (= Iterated Function System, see Chapter 5). Let

$$w_1(u) = +\sqrt{u - c} \quad \text{and} \quad w_2(u) = -\sqrt{u - c}.$$

be the transformations, which assign preimages of u, i.e. satisfy the equation $z^2 + c = u$. Then $\{w_n, p_n : n = 1, 2\}$ is an IFS (see Algorithm *JuliaIIM()*, Chapter 3), where $p_1 > 0$ and $p_2 > 0$ with $p_1 + p_2 = 1$ are probabilities which determine the random choice of w_1 and w_2. Unlike the transformations in Chapter 5 our transformations here are nonlinear. I.e. their contractiveness varies locally and for particular choices of c-values contraction may be so weak in some regions (e.g. close to parabolic fixed points) that the algorithm is inefficient.

(B) One computes the entire tree of iterated preimages. This leads rapidly into storage problems, if one uses the natural labeling as in Figure 4.4a, though it may be imbedded very elegantly and efficiently as a recursive procedure. For the whole tree one needs all 2^k preimages of the level k in order to compute level $k + 1$. If one, however, anticipates that N iterations suffice, then there is an obvious way to label the tree as in Figure 4.4b, which requires only $2(N - 1)$ (as compared to 2^{N-1}) units of storage.

4.2.1 The Mandelbrot set

The Mandelbrot set — discovered by B.B. Mandelbrot in 1980 [67] — is our next object. It is considered to be the most complex object mathematics has ever "seen". It is a rather peculiar fractal in that it combines aspects of self-similarity with the properties of infinite change. Subsequently we will discuss how the Mandelbrot set can be considered as the pictorial manifestation of order in the infinite variety of Julia sets (for each c there is a filled in Julia set K_c). The key of understanding the Mandelbrot set is the following rather crude classification: Each filled in Julia set K_c is[4]

- either connected (one piece)
- or a Cantor set (dust of infinitely many points)

(4.9)

We will give an explanation for (4.9), which will introduce a further algorithm for the K_c's. But before we discuss this classification we are now in a position to define the Mandelbrot set M:

$$M = \{c \in \mathbf{C} : K_c \text{ is connected}\} \tag{4.10}$$

---

[4] Figure 4.3a,b,c,e,g are connected Julia sets. A set is connected if it is not contained in the union of two disjoint and open sets. Figure 4.3d,f,h are Cantor sets. A Cantor set is in one-to-one correspondence to the set of all points in the interval [0,1], which have a triadic expansion (i.e. expansions with respect to base 3) in which the digit one does not occur.

MIIM - Modified Inverse Iteration Method : A detailed mathematical motivation is given in
([83], pp. 36-37). The idea of the algorithm is to make up for the non-uniform distribution of
the complete tree of iterated preimages by selecting an appropriate subtree which advances to
a much larger level of iteration k and forces preimages to hit sparse areas more often. The core
idea of the algorithm is this: Put J_c on a square lattice with small mesh size β. Then for any
box B of that mesh, stop using points from B for the preimage recursion, provided a certain
number N_{max} of such points in B have been used. Optimal choices of B and N_{max} depend
very much on J_c and other computergraphical parameters, such as the pixel resolution of the
given system (the mesh above should be thought of to be different from the pixel mesh of the
graphics device). Another variant, which also tries to determine a more efficient subtree in
Figure 4.4a than a typical IFS approach would yield, attempts to estimate the contractiveness
of w_1 and w_2 (see IIM). Given any point $u_{m_k} \neq u_0$ on the k-th level of the tree in Figure 4.4a
there is a unique path on the tree from u_{m_k} to u_0 which is determined by the forward iteration
of u_{m_k} (k times) :

$$f_c^k(u_{m_k}) = u_0 .$$

Now, the idea is to stop using u_{m_k} in the preimage recursion (i.e. to cut off the subtree starting
at u_{m_k}), provided that the derivative

$$|(f_c^k)'(u_{m_k})| = |\prod_{i=1}^{k} f_c'(u_{m_i})|, \quad m_0 = 0,$$

exceeds some bound D, which is the parameter of the algorithm. Here we have written

$$u_{m_i} = f_c^{k-i}(u_{m_k}), \quad i = 0, ..., k.$$

In other words, one tries to force the recursion to proceed on a subtree along which w_1 and
w_2 are less contractive. Of course, the above derivatives can be cheaply accumulated in the
course of the recursion :

$$\text{NewDerivative} = \gamma * \text{OldDerivative} * |u_{m_i}|,$$

where $\gamma = 2$. Moreover, γ can be used as an additional parameter to steer the method.

BSM - Boundary Scanning Method : Similar to MIIM, this method is based on a lattice - let's
assume a square lattice of mesh size β, which could be just the pixel lattice. Choose N_{max} – a
large integer – and R – a large number. Now let p be a typical pixel in the lattice with vertices
$v_i, i = 1, 2, 3, 4$. The algorithm consists in a labeling procedure for the v_i's :

$$v_i \text{ is labeled } 0, \quad \text{provided } v_i \in A_c(\infty), \quad \text{and}$$
$$v_i \text{ is labeled } 1, \quad \text{provided } v_i \in K_c.$$

Then p is called *completely labeled*, provided the vertices of p have labels which are *not all
the same*. A good approximation of J_c is obtained by coloring all completely labeled pixels in
the lattice. Thus it remains to decide (in approximation), whether $v_i \in A_c(\infty)$. The answer
is yes, provided that $|f_c^k(v_i)| > R$ for some $k \leq N_{max}$. Otherwise, it is assumed (with some
error depending on N_{max} and R, which define the 'iterative resolution' of the experiment)
that $v_i \in K_c$. Note that N_{max} should be chosen as a function of the pixel resolution of the
graphical device and the magnification factor, if one computes blow ups. Also, in some cases -
similar to IIM - no matter how large one may choose N_{max}, the resulting image will practically
not improve. The mathematical reasons for such effects are sometimes very delicate and are
exemplified and discussed in [83].

In other words, M lives in the complex c-plane, which serves as a plane of
"control parameters" for the K_c's. Figure 4.6 shows an image of M (in black).

MBSM - Modified Boundary Scanning Method : It is obvious that scanning all pixels of a lattice will be very time consuming, in particular for pixels p inside K_c. If J_c is connected, a much more economical algorithm is obtained in the following way. Assume that p_0 is a pixel in the lattice which is completely labeled. (p_0 could be obtained in various ways. For example, one could use the repelling fixed point of f_c to find p_0). Pixel p_0 is used as a seed for a *neighborhood search process*: Move all (immediately) neighboring pixels of p_0 onto a stack. Then test each pixel in the stack in three steps:

1. compute labels of vertices of a pixel from stack;

2. index whether pixel is completely labeled (where completely labeled is defined as in BSM);

3. if last pixel is completely labeled, push all those (immediate) neighbors onto stack, that have not been tested before.

For a more detailed description and further enhancements of this method we refer to [93].

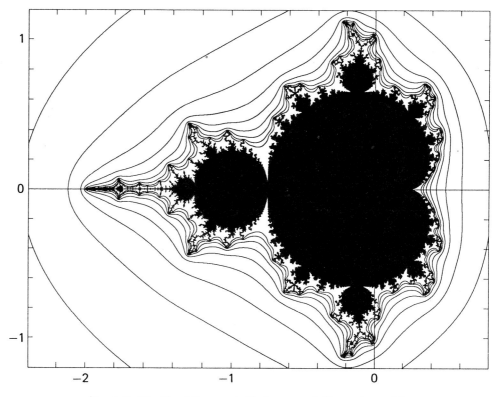

Fig. 4.6: The Mandelbrot set in black surrounded by equipotential curves.

How does one obtain pictures of M ? In other words, how does one distinguish K_c being connected from being a Cantor set in a computer? What is a computable indicator for the alternative (4.10)? We will see that as we explain why alternative (4.10) is in fact complete, we will also discover the desired indicator.

4.2.2   Hunting for K_c in the plane - the role of critical points

Suppose c is fixed and suppose we had to locate K_c. The following idea can be seen as a refinement of BSM. At the beginning, all we know is that K_c covers some finite piece of the z-plane. Thus, if we choose a very large disk centered at the origin of radius R and boundary S_R we can be sure that K_c is inside.

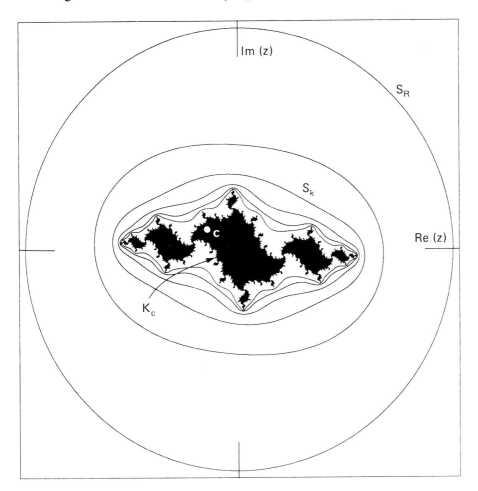

Fig. 4.7: Approximation of K_c by preimages of S_R: K_c connected

We shall now proceed recursively: For $k = 1, 2, \ldots$ we compute the preimage of S_{k-1} under f_c, i.e. with $S_0 = S_R$

$$S_k = \{ z : f_c(z) \in S_{k-1} \} \tag{4.11}$$

This amounts to solving equations $z^2 + c = b, b \in S_{k-1}$. Let us investigate the S_k's. We claim that S_1 is a closed curve which can be thought of as a deformed circle (see Figure 4.7) . This is certainly true if R is very large, because near $z = \infty$ (see Eqn. (4.2)) f_c acts like $f_0(z) = z^2$, slightly perturbed by c. Since the preimage of S_R under f_0 is a concentric circle with radius $\sqrt{R}$ we may conclude that S_1 is indeed a circle-like curve. Next we discuss the S_k's , $k > 1$. The following is a tautology:

- either all S_k's are circle-like
- or, there is $k^* \geq 1$, so that the S_k's are circle-like for   (4.12) $0 < k \leq k^*$ and they are not circle-like for $k > k^*$.

The nested sequence of circle-like S_k's in the first case approximates K_c better and better as $k \to \infty$ and, therefore, K_c must be connected. What can be said about the second case? Note that typically any $b \in S_{k-1}$ has two preimages a_1 and a_2 under f_c. The crucial observation is that for the preimages constituted by a_1 and a_2 only the following three cases are possible:

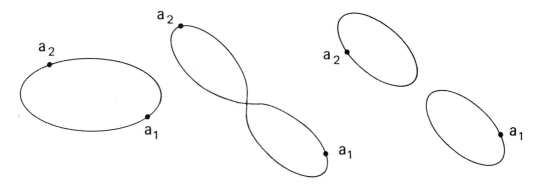

Fig. 4.8: Possible preimages of S_{k-1}

Either, as b turns around once on S_{k-1}, a_1 sweeps out one *halfcircle-like curve* (1) and a_2 sweeps out the other one, or they sweep out *two different curves* (3), or a_1 sweeps out the upper leaf and a_2 sweeps out the lower leaf of the *figure-eight* in (2). Why is it that the two circle-like curves obtained as a_1 and a_2 move around must be identical, separate or coincide in exactly one point? The answer leads to the heart of the matter: Assume, for example that a_1 and a_2 sweep out curves which intersect in two points w_1 and w_2 (see Figure 4.9).

Hence, w_1 is a double root of $z^2 + c = b_1$ and w_2 is a double root of $z^2 + c = b_2$ ($b_1 \neq b_2$, because otherwise one has a contradiction with f_c being of 2nd degree). But this is impossible, because the equation $z^2 + c = b$ allows double roots only when $b = c$ and those are $w_1 = 0 = w_2$. This justifies situation (2) in Figure 4.8 . One uses the following definition :

For $f_c(z) = z^2 + c$ c is called a *critical value*. Its preimage 0 is called a *critical point*.

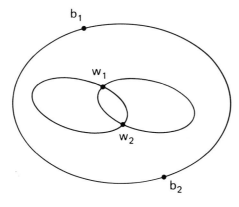

Fig. 4.9:

Thus, the second case in (4.12) is therefore completely characterized by either (2) or (3) in Figure 4.8 . That is, either $c \in S_{k^*}$, for some k^*, or c is between S_{k^*} and S_{k^*-1}. Figure 4.10 illustrates the second case in (4.12).

Observe that in Figure 4.7 we may conclude that $c \in K_c$, while in both cases of Figure 4.10 we have that $c \notin K_c$. In other words, we have just seen that

$$M = \{c : c \in K_c\} = \{c : 0 \in K_c\} = \{c : 0 \notin A_c(\infty)\}, \qquad (4.13)$$

and this is the computable indicator which led to early images of M, as e.g. Figure 4.16(a) or [67]. Also observe that since there is only one finite critical point[5] for f_c (using Eqn. (4.2) one shows easily that $f_c'(\infty) = 0$ as well) the preimage of a figure-eight will be a disjoint pair of figure-eights and the preimage of a circle-like curve inside the domain bounded by S_{k^*} will be a disjoint pair of circle-like curves, thus yielding nested sequences explaining the claimed Cantor set construction.

4.2.3   Level sets

The approximation of K_c by the S_k's has a very nice potential for graphical representations. ($\rightarrow$ LSM/J). The S_k's look like level sets of a profile in central projection from infinity. Figure 4.11 gives an example.

The contours of the level sets obtained by LSM - i.e. the S_k's discussed above, if T is the complement of a large disk - have a beautiful and important interpretation as equipotential curves of the electrostatic potential of K_c. This is our next objective which will lead to some more sophisticated algorithms.

---

[5]Note that for a general polynomial or more generally a rational function R one may have several critical points. They are obtained as zeros of $R'(z) = 0$, $z \in \mathbb{C} \cup \{\infty\}$.

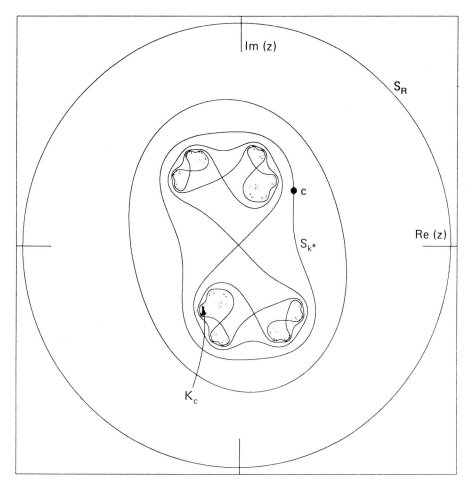

Fig. 4.10: Approximation of K_c by preimages of S_R : Cantor set construction

4.2.4   Equipotential curves

Let $c \in M$, i.e. K_c is connected. In that case one can show [29] that there is a one-to-one and onto biholomorphic[6] mapping $\phi_c : \mathbf{C} \setminus K_c \to \mathbf{C} \setminus D$, for which $\phi_c(z) \sim z$ when $z \sim \infty$ and which makes the following diagram commutative (where D is the closed unit disk)

$$
\begin{array}{ccc}
 & \phi_c & \\
\mathbf{C} \setminus D & \leftarrow & \mathbf{C} \setminus K_c \\
f_0 \downarrow & & \downarrow f_c \\
\mathbf{C} \setminus D & \leftarrow & \mathbf{C} \setminus K_c \\
 & \phi_c &
\end{array}
\qquad (4.14)
$$

---

[6] both ϕ_c and ϕ_c^{-1} are holomorphic i.e. complex analytic

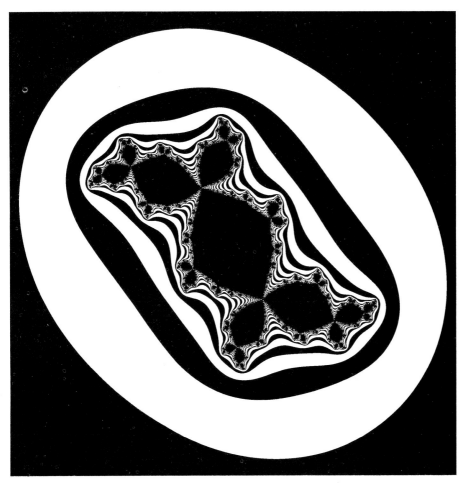

Fig. 4.11: A Julia set approximated by level sets

In other words, $f_0(z) = \phi_c(f_c(\phi_c^{-1}(z)))$, i.e. the iteration of f_0 outside of D
is equivalent to the iteration of f_c outside of K_c.

As an immediate consequence one has that

$$g_c(z) = \log(|\phi_c(z)|) \qquad (4.15)$$

is a *potential function* for K_c. Since the equipotential curves of D are just concentric circles one obtains the equipotential curves of K_c as images of these
circles under ϕ_c^{-1}. This fact confirms our heuristic argument in the construction
of the S_k's in Figure 4.7 . Together with g_c goes another natural function which
we interpret as the (continuous) escape time function

$$\varepsilon_c(z) = -\log_2(g_c(z)). \qquad (4.16)$$

(Note that as z approaches K_c then $\varepsilon_c(z)$ approaches ∞.) Now, what is the

Fig. 4.12: Escher-like tilings: Level sets of f_c with $c = 0$ around 0 with target set $T = K_c$

Fig. 4.13: Renderings of the potential in Eqs. (4.27) and (4.28). The top image shows the same region near the boundary of the Mandelbrot set as the cover image of the Scientific American, August 1985 (compare front cover). The bottom image displays a Cantor set.

LSM/J - Level Set Method : This algorithm is just a very powerful variant of BSM. We fix a square lattice of pixels, choose a large integer N_{max} (iteration resolution) and an arbitrary set T (target set) containing ∞, so that $K_c \subset \mathbf{C} \setminus T$. For example, $T = \{z : |z| \geq \frac{1}{\varepsilon}\}$, ε small, is a disk around ∞. Now we assign for each pixel p from the lattice an integer label $l_c(p; T)$ in the following way:

$$l_c(p; T) = \begin{cases} k, \text{ provided } f_c^i(p) \notin T \text{ and } f_c^k(p) \in T \text{ for } 0 \leq i < k \text{ and } k \leq N_{max} \\ 0, \text{ else.} \end{cases}$$

(writing $f_c(p)$ we mean, of course, $f_c(z)$ for some z representing p, as for example its center.) The interpretation of a nonzero $l_c(p; T)$ is obvious: p escapes to ∞ and $l_c(p; T)$ is the "escape time" - measured in the number of iterations - needed to hit the target set around ∞. The collection of points of a fixed label, say k, constitutes a level set, the boundary of which is just the union of S_{k-1} and S_k, provided T is the complement of a large disk. Since T can essentially be anything, this method has a tremendous artistic potential. For example, T could be a so-called p-norm disk ($0 < p < \infty$)

$$T = \{z : (|Re(z)|^p + |Im(z)|^p)^{\frac{1}{p}} \geq \frac{1}{\varepsilon}\},$$

or a scaled filled-in Julia set or something designed by hand (see Figure 4.12). This method opens a simple and systematic approach to Escher-like tilings.

---

relation between $\varepsilon_c(z)$ and the contours defined by LSM, where

$$T = \{z : |z| \geq \frac{1}{\delta}\}$$

is the complement of a large disk ? It is known, see [77], that

$$|\varepsilon_c(z) - l_c(z; T)| \text{ is uniformly bounded for } z \notin K_c, \ z \in \mathbf{C} \setminus T \quad (4.17)$$

In other words, the contours obtained by LSM are approximations of the equipotential curves of K_c.

The above considerations use in an essential way that K_c is connected, because otherwise ϕ_c does not exist as in Eqn. (4.14). If, however, K_c is a Cantor set, i.e. $c \notin M$, then ϕ_c exists at least outside S_k^* (see Figure 4.10) and there the relation $f_0(z) = \phi_c(f_c(\phi_c^{-1}(z)))$ is still true. Thus, again, the contours obtained by LSM far from K_c will be circle-like, in agreement with our previous discussion. In particular, one has that in that case

$$c \notin M : \phi_c(c) \text{ is still defined}. \quad (4.18)$$

An important representation of the potential $g_c(z)$ for graphical and mathematical purposes is given by (see [29]):

$$g_c(z_0) = \lim_{k \to \infty} \frac{\log |z_k|}{2^k}, \quad (4.19)$$

where

$$z_k = z_{k-1}^2 + c, \ k = 1, 2, \dots .$$

```
ALGORITHM MSetLSM  (MSet, nx, ny, xmin, xmax, ymin, ymax, maxiter)
Title           Mandelbrot set via Level Set Method (LSM)

Arguments    MSet[][]     output array of integer type, size nx by ny
             nx, ny       image resolution in x- and y-direction
             xmin, xmax   low and high x-value of image window
             ymin, ymax   low and high y-value of image window
                          aspect ratio of window is : nx by ny
             maxiter      maximal number of iterations
Variables    ix, iy       integer
             cx, cy       real
Functions    MSetLevel()  returns level set of a point

BEGIN
    FOR iy = 0 TO ny-1 DO
        cy := ymin + iy * (ymax - ymin) / (ny - 1)
        FOR ix = 0 TO nx-1 DO
            cx := xmin + ix * (xmax - xmin) / (nx - 1)
            MSet[ix][iy] := MSetLevel (cx, cy, maxiter)
        END FOR
    END FOR
END
```

```
ALGORITHM MSetLevel  (cx, cy, maxiter)
Title           Function returning level set of a point

Arguments    cx, cy       point to be tested
             maxiter      maximal number of iterations
Variables    iter         integer
             x, y, x2, y2 point coordinates and squares
             temp         real scratch variable

BEGIN
    x := y := x2 := y2 := 0.0
    iter := 0
    WHILE (iter < maxiter) AND (x2 + y2 < 10000.0) DO
        temp := x2 - y2 + cx
        y := 2 * x * y + cy
        x := temp
        x2 := x * x
        y2 := y * y
        iter := iter + 1
    END WHILE
    RETURN (iter)
END
```

Convergence is very rapid, once $|z_k|$ is large[7]. Using this information A. Douady and J. Hubbard [29] go on to show that there is in fact a one-to-one and onto biholomorphic mapping

$$\Phi : \mathbf{C} \setminus M \to \mathbf{C} \setminus D \qquad (4.20)$$

$D = \{z : |z| \leq 1\}$, which is given by $\Phi(c) = \phi_c(c)$, see (4.18). That provides a potential function for M

$$G(c) = \log |\Phi(c)|, \qquad (4.21)$$

which, similar to Eqn. (4.19) can be represented by

$$G(c) = \lim_{k \to \infty} \frac{\log |z_k|}{2^k}, \qquad (4.22)$$

where

$$z_k = z_{k-1}^2 + c, \ z_0 = c, \ k = 1, 2, \ldots$$

where again convergence is rapid for large $|z_k|$. The existence of Φ above means that all equipotential curves of M are circle-like, i.e. images of concentric circles around D under Φ^{-1}. An immediate consequence of this is that

$$M \ \ \text{is connected}.$$

This appears to be a truly remarkable result in view of the complexity of structures seen at the boundary of M (see [83] and Figure 4.21). The potential G leads to a *(continuous) escape time function*

$$E(c) = -\log_2(G(c)), \qquad (4.23)$$

which is appropriate to be compared with a *(discrete) escape time function* (analogous to Eqn. (4.16)):

$$L(c; T) = \begin{cases} k, & \text{if } f_c^i(0) \notin T \text{ and } f_c^k(0) \in T \text{ for } 0 \leq i < k \\ 0, & \text{else} \end{cases} \qquad (4.24)$$

(T is a target set containing ∞.) If T is given by $T = \{z : |z| > \frac{1}{\varepsilon}\}$, then

$$|E(c) - L(c; T)| \text{ is uniformly bounded for } |c| < \frac{1}{\varepsilon} \text{ and } c \notin M, \ (4.25)$$

---

[7]Let us understand why Eqn. (4.19) is true. Combining Eqn. (4.14) and Eqn. (4.15) yields:

$$g_c(f_c(z)) = \log |\phi_c(f_c(z))| = \log |(\phi_c(z))^2| = 2g_c(z).$$

Thus, iterating the argument one obtains:

$$g_c(z) = \frac{g_c(f_c^k(z))}{2^k} = \frac{\log |\phi_c(f_c^k(z))|}{2^k}.$$

Finally, as $k \to \infty$ we have that $f_c^k(z) \to \infty$ as well, and, thus, Eqn. (4.19) follows from the property that $\phi_c(z) \sim z$ when $z \sim \infty$.

i.e. the contours defined by L outside of M are approximations of equipotential curves for the electrostatic potential of M. The above results about the potential of K_c and M give rise to several very interesting algorithms (LSM/M,CPM/J, CPM/M).

LSM/M - Level Set Methods for M : This algorithm is strictly analogous to LSM/J. We fix a square lattice of pixels, choose a large integer N_{max} (iteration resolution) and an arbitrary set T (target set) containing ∞ , so that $M \subset \mathbf{C} \setminus T$. Modify Eqn. (4.24) by:

$$L(c;T) = \begin{cases} k, & \text{if } f_c^i(0) \notin T \text{ and } f_c^k(0) \in T \text{ for } 0 \leq i < k \text{ and } 1 \leq k \leq N_{max} \\ 0, & \text{else} \end{cases}$$

$$(4.26)$$

Again it might be useful to identify a certain number of level sets near M to obtain reasonable pictures. Figure 4.14 shows a result of this method and a blow up near the boundary of M, where alternating levels are colored black and white. The boundary of the levels in Figure 4.14 can be interpreted as equipotential curves according to Eqn. (4.25) , see Figure 4.6 .

CPM/J - Continuous Potential Method for Julia Sets : This algorithm is based on the representation Eqn. (4.19) and allows to represent the potential of K_c as a smooth parametrized surface $pot_c : \mathbf{C} \setminus K_c \to \mathbf{C} \times \mathbf{R}$, $\mathbf{R}$ denotes the real numbers, (see Plate 27) given by the graph of $g_c(z)$, which is approximately given by

$$pot_c(z_0) = \left(z_0, \frac{\log(|z_n|)}{2^n} \right),$$

$$(4.27)$$

where

$$z_k = z_{k-1}^2 + c, \ k = 1, 2, ..., n, \ n = l_c(z_0; T)$$

and

$$T = \{z : |z| \geq \frac{1}{\varepsilon}\}, \ \varepsilon \text{ small}.$$

For 3D renderings of pot_c it is usually appropriate to transform the potential to make up for too slow or too steep descents, for example. In Plate 27, one has a rendering of a function like $1 - (g_c(z))^\alpha$, with $\alpha = \frac{1}{2}$ together with a cut-off function for z very close to K_c (see Section 2.7). We have used this method successfully for both connected and Cantor set K_c's (see Figure 4.13). Note that the level curves of pot_c are circle-like far outside.

CPM/M - Continuous Potential Method for Mandelbrot Set : This algorithm is strictly analogous to CPM/J in using (4.22) for a smooth parametrized surface $Pot : \mathbf{C} \setminus M \to \mathbf{C} \times \mathbf{R}$ (see Plates 25, 26, 28 and Figure 4.13) of the potential given by the graph of G, which is approximately given by

$$Pot(c) = \left(c, \frac{\log(|z_n|)}{2^n} \right),$$

$$(4.28)$$

where

$$z_k = z_{k-1}^2 + c, \ z_0 = c, \ k = 1, 2, ..., n, \ n = L(c; T)$$

and

$$T = \{z : |z| \geq \frac{1}{\varepsilon}\}, \ \varepsilon \text{ small}.$$

Note that all level curves of Pot are circle-like, even close to M.

```
ALGORITHM MSetCPM (MSet, nx, ny, xmin, xmax, ymin, ymax, maxiter)
Title           Mandelbrot set via Continuous Potential Method (CPM)

Arguments    MSet[][]      output array of real type, size nx by ny
             nx, ny        image resolution in x- and y-direction
             xmin, xmax    low and high x-value of image window
             ymin, ymax    low and high y-value of image window
                           aspect ratio of window is : nx by ny
             maxiter       maximal number of iterations
Variables    ix, iy        integer
             cx, cy        real
Functions    MSetPot()     returns potential of a point

BEGIN
     FOR iy = 0 TO ny-1 DO
          cy := ymin + iy * (ymax - ymin) / (ny - 1)
          FOR ix = 0 TO nx-1 DO
               cx := xmin + ix * (xmax - xmin) / (nx - 1)
               MSet[ix][iy] := MSetPot (cx, cy, maxiter)
          END FOR
     END FOR
END
```

```
ALGORITHM MSetPot (cx, cy, maxiter)
Title           Function returning potential of a point

Arguments    cx, cy        point to be tested
             maxiter       maximal number of iterations
Variables    iter          integer
             x, y, x2, y2  point coordinates and squares
             temp          real scratch variable
             potential     real variable, potential

BEGIN
     x := cx; x2 := x * x
     y := cy; y2 := y * y
     iter := 0
     WHILE (iter < maxiter) AND (x2 + y2 < 10000.0) DO
          temp := x2 - y2 + cx
          y := 2 * x * y + cy
          x := temp
          x2 := x * x
          y2 := y * y
          iter := iter + 1
     END WHILE
     IF (iter < maxiter) THEN
          potential := 0.5 * log (x2 + y2) / power (2.0, iter)
     ELSE
          potential := 0.0
     END IF
     RETURN (potential)
END
```

4.2.5   Distance estimators

Another useful property is due to J. Milnor and W. Thurston [77] : Using the potential function $G(c)$ one can show that for c outside of M the distance $d(c, M)$ from c to the boundary of M can be estimated as

$$d(c, M) \leq \frac{2 \, \sinh \, G(c)}{|G'(c)|}, \qquad (4.29)$$

where $G'(c)$ denotes the usual gradient of G. DEM/M[8] is an algorithm which is based on Eqn. (4.29) and indicates how to compute the boundary of M. A reasonable and practical approximation to the right hand side of Eqn. (4.29) is obtained in the following way : For c near M we have that $G(c)$ is close to 0, thus we may approximate $\sinh(G(c))$ by $G(c)$. To approximate $G(c)$ and $|G'(c)|$ we make use of Eqn. (4.22) and the Cauchy-Riemann differential equations to obtain:

$$\frac{\sinh \, G(c)}{|G'(c)|} \approx \frac{|z_n|}{|z_n'|} \log |z_n|, \qquad (4.30)$$

where

$$z_n' = \frac{dz_n}{dc}.$$

The distance estimation for points c near the Mandelbrot set applies also for z near K_c (connected). DEM/J is an algorithm analogous to DEM/M (see Figure 4.15b). Though being rigorously justified only for connected K_c we have used it with very nice success also for the Cantor set case such as in Figure 4.15.

4.2.6   External angles and binary decompositions

The potential of M is most crucial for an understanding of M. While we have discussed equipotential curves already from various points of view we will now briefly sketch a discussion of field lines. These are orthogonal to the equipotential curves and can be formally defined to be the images of the field lines of the potential of $D = \{z : |z| \leq 1\}$ under Φ^{-1} (see Figure 4.17) :

$$L_\alpha = \left\{ \Phi^{-1}(r \cdot e^{2\pi i \alpha}) : r > 1 \right\}$$

---

[8] Our description here is also suitable for coloring the exterior of M (Plate 28 was obtained in this way). Simply use Eqn. (4.31). This estimate, however, can also be used to set up extremely fast algorithms to obtain images of M and its blow ups: Cover the exterior of M by disks whose radii are determined by this equation. Another alternative would be to use Eqn. (4.31) for a neighborhood search process, similar to MBSM.

Fig. 4.14: Equipotential curves of M ; LSM/M

Fig. 4.15: $K_c, c = -0.74543 + 0.11301i$, obtained by MIIM (a) and DEM/J (b)

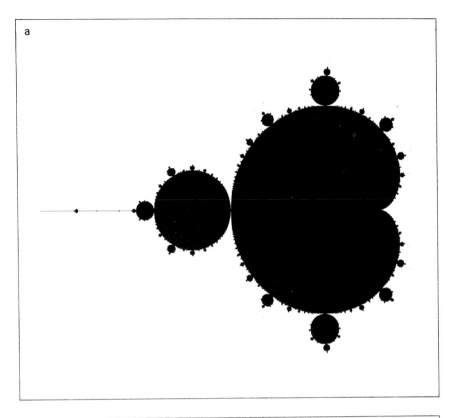

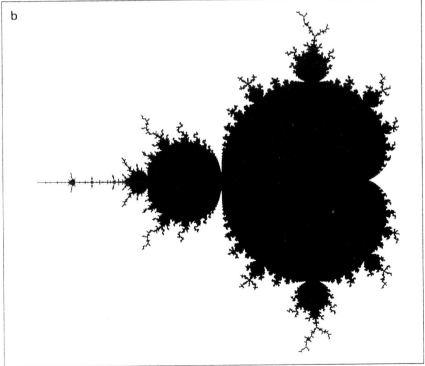

Fig. 4.16: *M* obtained by BSM (a) and DEM/M (b)

DEM/M - Distance Estimator Method for Mandelbrot set : Choose N_{max} (maximal number of iterations) and $R(= \frac{1}{\varepsilon})$ where $T = \{z : |z| \leq \frac{1}{\varepsilon}\}$ is the target set around ∞. For each c we will determine a label $l(c)$ from $\{0, \pm 1, 2\}$ (0 for $c \in M$, $\{+1, -1\}$ for c close to M, 2 for c not close to M) : Compute

$$z_{k+1} = z_k^2 + c, \; z_0 = 0, k = 0, 1, 2, ...$$

until either $|z_{k+1}| \geq R$ or $k = N_{max}$. In the second case($k = N_{max}$) we set $l(c) = 0$. In the other case we have $|z_n| \geq R$ with $n = k + 1 = L(c; T)$ and c is still a candidate for a point close to M. Thus we try to estimate its distance, having saved the orbit $\{z_0, z_1, ..., z_n\}$:

$$z'_{k+1} = 2 z_k z'_k + 1, \; z'_0 = 0, \; k = 0, 1, ..., n - 1$$

If in the course of the iteration of this equation we get an overflow, i.e. if

$$|z'_{k+1}| \geq \; \text{OVERFLOW}$$

for some k then c should be very close to M, thus we label c by -1. If no overflow occurred, then we estimate the distance of c from M by

$$Dist(c, M) = 2 \frac{|z_n|}{|z'_n|} \log |z_n| \qquad\qquad (4.31)$$

and set

$$l(c) = \begin{cases} 1, \text{ if } Dist(c, M) < \text{DELTA} \\ 2, \text{ otherwise} \end{cases}$$

It turns out that the images depend very sensitively on the various choices (R, N_{max}, OVER-FLOW, DELTA and blow-up factor when magnifying M). Plotting all pixels which represent c-values with $|l(c)| = 1$ provides a picture of the boundary of M. Figure 4.21 was obtained in this way. Figure 4.16 shows a comparison between BSM and DEM/M. The estimate value of $d(c, M)$ given by $Dist(c, M)$ can be used to define a surface

$$\begin{aligned} \Delta : \mathbf{C} \setminus \{c : l(c) = 0 \text{ or } l(c) = -1\} \;\; &\to \;\; \mathbf{C} \times \mathbf{R} \\ c \;\; &\mapsto \;\; (c, Dist(c, M)) \end{aligned}$$

Sometimes it is convenient, similar to CPM/M, to rescale Δ or to invert the surface, i.e. for example $\Delta^*(c) = (c, 1 - Dist(c, M))$. Note that the level curves of Δ , unlike for Pot in CPM/M, will not be circle-like near the boundary of M. Plate 29 shows a 3D rendering of the distance estimate, using the techniques described in Section 2.7.

One uses the angle α of a field line for D as a convenient index for field lines of M. One of the most prominent open questions about M is whether field lines actually "land" on the boundary of M, i.e. formally, whether the following limit exists:

$$\lim_{r \to 1} \Phi^{-1}(r \cdot e^{2\pi i \alpha}).$$

The answer would be yes (as a consequence of a famous result due to Caratheo-dory), if one would only know that M being connected is also *locally con-nected*.[9] Douady and Hubbard have shown that the limit exists if $\alpha = \frac{p}{q}$ is

---

[9] M would be locally connected, provided for every $c \in M$ and every neighborhood U of c there is a connected neighborhood V in U. Figure 4.18 is an example of a set which is connected but *not* locally connected

ALGORITHM **MSetDEM**	(MSet, nx, ny, xmin, xmax, ymin, ymax, maxiter, threshold)	
Title	Mandelbrot set via distance estimate (DEM)	

Arguments	MSet[][]	output array of boolean type, size nx by ny
	nx, ny	image resolution in x- and y-direction
	xmin, xmax	low and high x-value of image window
	ymin, ymax	low and high y-value of image window
		aspect ratio of window is : nx by ny
	maxiter	maximal number of iterations
	threshold	critical distance from MSet in pixel units
Variables	ix, iy	integer
	cx, cy	real
	dist	distance estimate
Functions	MSetDist()	returns distance estimate of a point to the MSet

```
BEGIN
    delta := threshold * (xmax - xmin) / (nx - 1)
    FOR iy = 0 TO ny-1 DO
        cy := ymin + iy * (ymax - ymin) / (ny - 1)
        FOR ix = 0 TO nx-1 DO
            cx := xmin + ix * (xmax - xmin) / (nx - 1)
            dist := MSetDist (cx, cy, maxiter)
            IF dist < delta THEN
                MSet[ix][iy] := 1
            ELSE
                MSet[ix][iy] := 0
            END IF
        END FOR
    END FOR
END
```

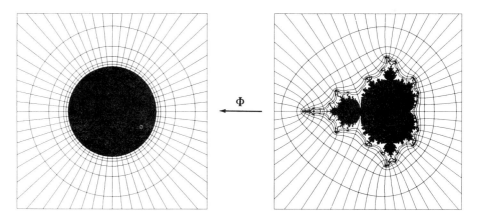

Fig. 4.17: Equipotential and field lines of M

a rational number. A nice way to visualize the L_α's is provided by an experiment which we call binary decomposition of level sets (see BDM/J, BDM/M). For details of the mathematical background we refer to [83], pp. 40-44, 64-76. Figure 4.19 shows binary decompositions for three K_c's and one observes that

ALGORITHM **MSetDist** (cx, cy, maxiter)		
Title		Distance estimator for points near Mandelbrot set
Arguments	cx, cy	point to be tested
	maxiter	maximal number of iterations
Variables	iter, i	integers
	x, y, x2, y2	point coordinates and squares
	temp	real scratch variable
	xder, yder	derivative, real and imaginary parts
	xorbit[]	array of real parts of orbit, size > maxiter
	yorbit[]	array of imaginary parts of orbit, size > maxiter
	dist	real, distance estimate
	huge	large number used for stopping the iteration
	flag	boolean, indicates overflow in derivative calculation
Constants	overflow	maximal size of derivative

```
BEGIN
    x := y := x2 := y2 := dist := xorbit[0] := yorbit[0] := 0.0
    iter := 0
    huge := 100000.0
    WHILE (iter < maxiter) AND (x2 + y2 < huge) DO
        temp := x2 - y2 + cx
        y := 2 * x * y + cy
        x := temp
        x2 := x * x
        y2 := y * y
        iter := iter + 1
        xorbit[iter] := x
        yorbit[iter] := y
    END WHILE
    IF x2 + y2 > huge THEN
        xder := yder := 0.0
        i := 0
        flag := FALSE
        WHILE (i < iter) AND (NOT flag) DO
            temp := 2 * (xorbit[i] * xder - yorbit[i] * yder) + 1
            yder := 2 * (yorbit[i] * xder + xorbit[i] * yder)
            xder := temp
            flag := max (abs (xder), abs(yder)) > overflow
            i := i + 1
        END DO
        IF (NOT flag) THEN
            dist := log (x2 + y2) * sqrt (x2 + y2) / sqrt (xder * xder + yder * yder)
        END IF
    END IF
    RETURN (dist)
END
```

angles of the form $\frac{p}{2^n} \cdot 2\pi$ are pronounced. The general idea is to subdivide the target set in the LSM/J (resp. LSM/M) algorithm into two parts, and then label according to the alternative whether $z_n, n = l_c(z_0; T)$ (resp. $n = L(c; T)$), hits the first or second part.

DEM/J - Distance Estimator Method for Julia Sets : Let c be fixed. Choose N_{max} (maximal number of iterations) and $R = \frac{1}{\varepsilon}$, where $T = \{z : |z| \geq \frac{1}{\varepsilon}\}$ is the target set around ∞. For each z_0 we will determine a label $l(z_0)$ from $\{0, \pm 1, 2\}$ (0 for $z_0 \in K_c$, $\{+1, -1\}$ for z_0 close to K_c, 2 for c not close to K_c) : Compute

$$z_{k+1} = z_k^2 + c, \ k = 0, 1, 2, \ldots$$

until either $|z_{k+1}| \geq R$ or $k = N_{max}$. In the second case($k = N_{max}$ we set $l(z_0) = 0$. In the other case we have $|z_n| \geq R$ with $n = k + 1 = l_c(z_0; T)$ and z_0 is still a candidate for a point close to K_c. Thus we try to estimate its distance, having saved the orbit $\{z_0, z_1, \ldots, z_n\}$:

$$z_{k+1}' = 2 z_k z_k', \ z_0' = 1, \ k = 0, 1, \ldots, n-1 \tag{4.32}$$

If in the course of the iteration of Eqn. (4.32) we get an overflow, i.e. if

$$|z_{k+1}'| \geq \text{ OVERFLOW}$$

for some k then z_0 should be very close to K_c, thus we label z_0 by -1. If no overflow occurred, then we estimate the distance of z_0 from K_c by

$$dist(z_0, K_c) = 2 \frac{|z_n|}{|z_n'|} \log |z_n| \tag{4.33}$$

and set

$$l(c) = \begin{cases} 1, & \text{if } dist(z_0, K_c) < \text{DELTA} \\ 2, & \text{otherwise} \end{cases}$$

Figure 4.15 shows a result of a comparison between DEM/J and MIIM.

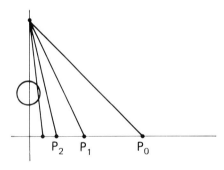

Fig. 4.18: Example of non-locally connected set.(Let $P_n = (\frac{1}{2^n}, 0) \in \mathbf{R}^2$, take lines from (0,1) to P_n and also from (0,1) to (0,0) ; the collection of these line segments is not locally connected.)

4.2.7 Mandelbrot set as one-page-dictionary of Julia sets

Our last aspect of the Mandelbrot set deals with one of its most phenomenal properties. In a certain sense it can be considered to be a one-page-dictionary of all Julia sets. This fact immediately gives a flavor of its unimaginable complexity and it also implies that the Mandelbrot set is not self-similar because it stores infinitely many different Julia sets.

More precisely, if one looks at M with a microscope focused at c, what one sees resembles very much what one sees if one looks at K_c with the same micro-

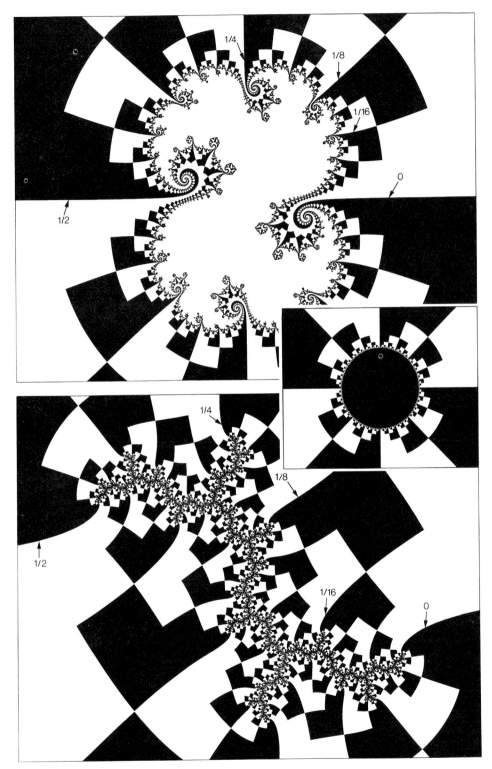

Fig. 4.19: Binary decompositions for three K_c's : $c = 0$ center, $c = 0.32 + 0.043\,i$ above, $c = i$ below; obtained by BDM/J

Fig. 4.20: Binary decomposition of M-set in the $\frac{1}{c}$-plane, rotated.

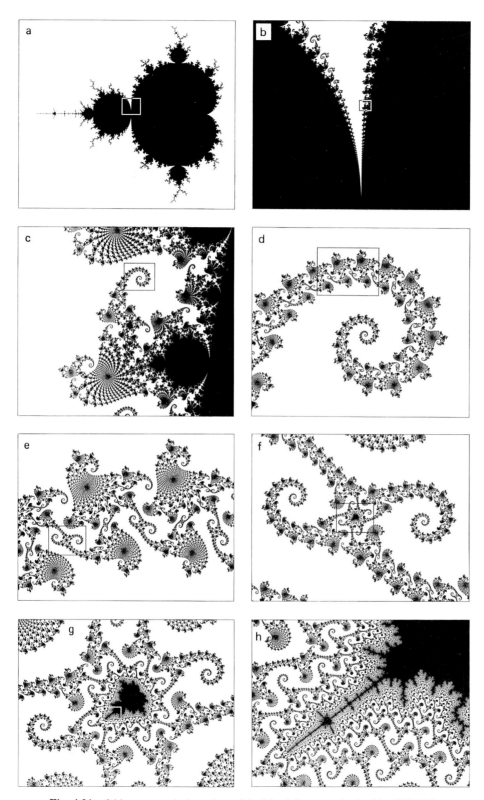

Fig. 4.21: 8 blow ups at the boundary of the Mandelbrot set obtained by DEM/M

BDM/J - Binary Decomposition Method for Julia Sets (with respect to a fixed angle α_0): This algorithm is a variant of LSM/J. We assume that $T = \{z : |z| \geq \frac{1}{\epsilon}\}$ and that each pixel has a label, i.e. $l_c(z_0; T)$ is computed, i.e. the (discrete) escape time for z_0 outside of K_c is known. We now extend our labeling function by determining a second component

$$
\begin{aligned}
l_c^*(z_0; T) &= (k, m) \text{ where} \\
k &= l_c(z_0; T) \text{ and} \\
m &= \begin{cases} 0, & \text{if } 2^k \alpha_0 \leq \arg(z_k) \leq 2^k \alpha_0 + \pi (\mod 2\pi) \\ 1, & \text{otherwise} \end{cases}
\end{aligned}
\tag{4.34}
$$

Here α_0 is any angle between 0 and 2π and $\arg(z_k) = \beta$, if $z_k = re^{2\pi i \beta}$. In Figure 4.19 we chose $\alpha_0 = 0$. The second component can be used as an additional color information (see [83], p. 91) or as in Figure 4.19, where the first component is simply disregarded and 0 is coded white and 1 is coded black for the second component.

BDM/M - Binary Decomposition Method for the M-set : This method is strictly analogous using $L(c; T)$ in LSM/M. Figure 4.20 shows the result in the $\frac{1}{c}$-plane.

scope still focused at c, and the resemblance tends to become perfect (except for a change in scale) when one increases the magnifying power. We illustrate this marvelous "image compression" property in Figure 4.22 . A region along the cardioid (see top of Figure 4.22) is continuously blown up and stretched out, so that the respective segment of the cardioid becomes a line segment[10]. The result is shown in the center of Figure 4.22. Attached to each of the disk-components there one observes dendrite structures with $3, 4, ..., 8$ branches (from left to right). Now we choose as particular c-values the major branch points of these dendrites (see arrows). For each of these 6 c-values we have computed the associated Julia sets (using DEM/J) which are displayed in the bottom of Figure 4.22. We observe a striking resemblance in the structures and the combinatorics of the dendrite structures. Figure 4.23 is yet another example. There we show a tiny window at the boundary of M blown up by a factor of approximately 10^6 (top part) and compare it with a blow up (factor $\sim 10^6$) of an associated Julia set (bottom part).

For all other properties of the Mandelbrot set, for example, the structure in its interior we refer to [83].

---

[10] The blow-up is obtained in the following way: Attached to the cardioid of M one observes an infinity of disk-like components. Each of these characterizes certain periodic cycles of f_c. For example, the cardioid itself collects all c-values for which f_c has an attracting fixed point; the large disk attached to the cardioid collects all c-values for which f_c has an attractive 2-cycle; the two next larger disks attached to the cardioid collect all c-values for which f_c has an attracting 3-cycle; etc. The conjecture that the size of a disk characterizing a p-cycle is proportional to $\frac{1}{p^2}$ was recently proven by J. Guckenheimer and R. McGehee [47]. Our blow-up factor is chosen accordingly to the result that all disks in Figure 4.22 (center) have the same size. They characterize — from left to right — attracting 3-, 4-, ..., 8-cycles.

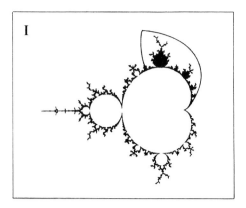

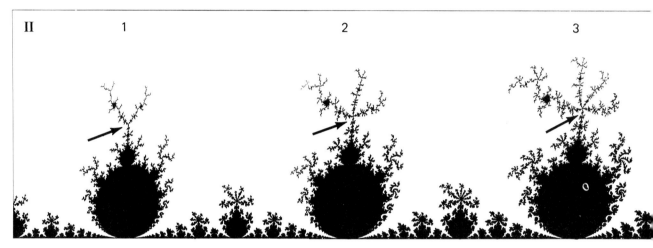

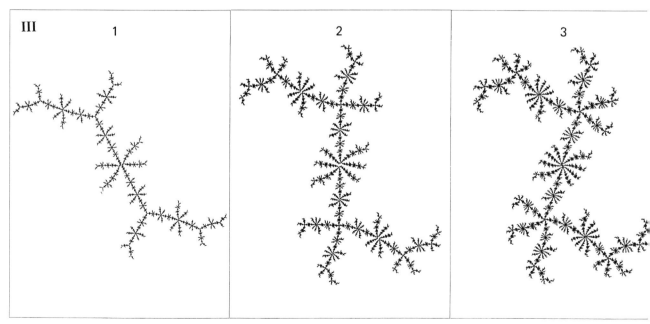

Fig. 4.22: Correspondence between Mandelbrot set and Julia sets; images obtained by DEM/J and DEM/M

Fig. 4.23: "Image compression" in the Mandelbrot set demonstrated for the c-value $-0.745429 + 0.113008i$. The top left shows a close up view of the Mandelbrot set around this c-value. The vertical size of the window is 0.000060. The Julia set for the c-value is shown in the bottom right image. The close up view of this Julia set in the bottom left is centered around the same above c-value. It reveals the same double spiral as in the picture above . The vertical window size is 0.000433, and the Julia set has been rotated by 55° counter clockwise around the center of the window.

4.3   Generalizations and extensions

Most of the algorithms introduced in Section 4.2 for $f_c(z) = z^2 + c$ have natural extensions to general rational mappings of the complex plane (for extensions to other maps like, for example, transcendental maps see Chapter 3). So let $R(z)$ be a general rational mapping, i.e.

$$R(z) = \frac{p(z)}{q(z)}, \qquad (4.35)$$

where both $p(z)$ and $q(z)$ are polynomials of degree d_p and d_q without a common divisor. Then the degree of R is $d_R = \max\{dp, dq\}$. Let's assume that $d_R \geq 2$. First of all, the definition of the Julia set J_R of R is literally the same as that in Eqn. (4.7) or Eqn. (4.8), i.e. the algorithms based on these characterizations - like IIM and MIIM - are exactly the same. In general, however, the role of ∞ is a different one. For example, if

$$R(z) = \left(\frac{z-2}{z}\right)^2 \qquad (4.36)$$

then $\infty \mapsto 1 \mapsto 1$ and $R'(1) = -4$, i.e. 1 is a repelling fixed point and ∞ is one of its two preimages. Actually, Eqn. (4.36) has been a very important example in the history of Julia sets: It is one of the exotic examples for which $J_R = \mathbf{C} \cup \{\infty\}$, a situation which never appears in the family f_c being discussed in Section 4.2.

4.3.1   Newton's Method

Another famous example is

$$R(z) = \frac{2z^3 + 1}{3z^2} \qquad (4.37)$$

which the reader will immediately recognize, if we rewrite it as

$$R(z) = z - \frac{z^3 - 1}{3z^2}$$

i.e. Eqn. (4.37) is the familiar Newton method to solve the equation $z^3 - 1 = 0$. In fact the problem to understand Eqn. (4.37) goes back to Lord A. Cayley of Cambridge (1879) and has been one of the guiding problems for Julia and Fatou to develop their beautiful theory (see [84] for a historic discussion). Note that the 3 zeros of $z^3 - 1 = 0$ (being $z_k = exp(\frac{2\pi i k}{3})$, $k = 1, 2, 3$) are attracting fixed points of R constituting 3 basins of attraction $A(z_k)$, $k = 1, 2, 3$. The amazing fact is that

$$J_R = \partial A(z_k), \ k = 1, 2, 3 \qquad (4.38)$$

i.e. $\partial A(z_1) = \partial A(z_2) = \partial A(z_3)$, which means that J_R is a set of triple-points with respect to the three basins. It turns out that this property seen in Figure 4.24 is in fact a general property:

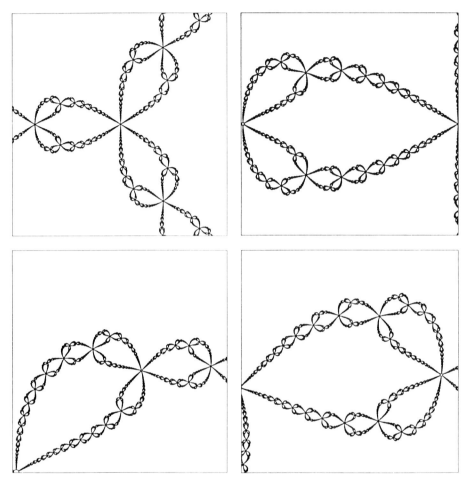

Fig. 4.24: Julia set of $R(z) = (2z^3 + 1)/(3z^2)$ and blow ups; a set of triple-points

Assume that R is a rational mapping with attracting fixed points (periodic points) $z_1, ..., z_k$ and associated basins of attraction $A(z_i), i = 1, ..., k$. Then

$$J_R = \partial A(z_i), \quad i = 1, ..., k. \qquad (4.39)$$

In other words, one may have quadruple -points, quintuple-points, ... , etc. Examples are also easy to generate. Just choose a polynomial with a certain number of zeros and apply Newton's method. Note that for Eqn. (4.37) we have that $R(\infty) = \infty$. However, here ∞ is a repelling fixed point. This is true for Newton's method applied to any polynomial of degree ≥ 2. Thus, for rational mappings obtained in this way one has that

$$J_R = \text{ closure } \{z : R_k(z) = \infty \text{ for some k}\}.$$

Obviously, for a numerical interpretation one does not like to work with ∞. Therefore, one computes one of the finite preimages of ∞ under R and rather works with that , for example, for Eqn. (4.37) one has $0 \mapsto \infty \mapsto \infty$, i.e.

$$J_R = \text{ closure } \{z : R_k(z) = 0 \text{ for some k} \}.$$

Property (4.39) has an impact on BSM, MBSM, LSM/J and BDM/J. In fact they translate literally by just replacing the role of ∞ with that of any of the zeros of the polynomial under consideration for Newton's method, say z_1 and z_2. Then, rather then testing orbits going to ∞ versus orbits remaining bounded, one tests simply orbits going to z_1 versus going to z_2. In fact, Figure 4.24 was obtained by BSM for the mapping (4.37) and Figure 4.25b is a result of BDM/J.

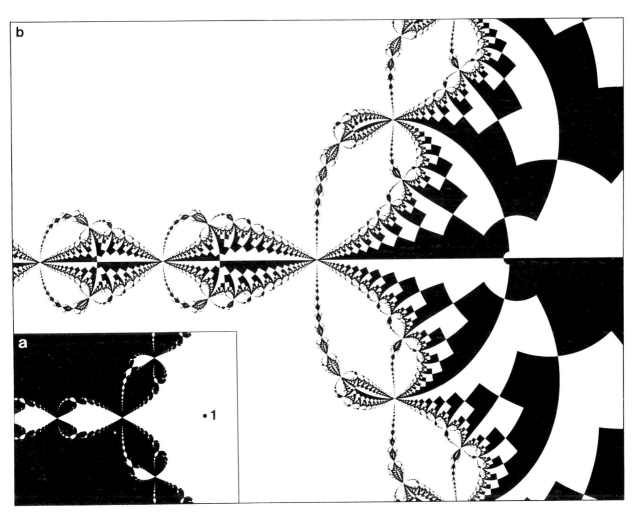

Fig. 4.25: **a**. Basin $A(1)$ of mapping (4.37) in white; BSM. **b**. Binary decomposition of $A(1)$; BDM/J

4.3.2   Sullivan classification

One of the deeper mathematical questions about rational functions R is to characterize $\mathbf{C} \cup \{\infty\} \backslash J_R$, the complement of the Julia set. Recently, D. Sullivan [98] has solved this problem by showing that there are only four different cases possible, i.e. each component of the complement of J_R is related to (see [83] for details and images):

1)   attractive case (e.g. $R(z) = z$ and $|R'(z)| < 1$),
2)   parabolic case (e.g. $R(z) = z$ and $R'(z) = \exp(2\pi i\alpha)$
      and $\alpha = \frac{p}{q}$ (rational)),
3)   Siegel-disk case (e.g. $R(z) = z$ and $R'(z) = \exp(2\pi i\alpha)$ $\left.\begin{array}{c}\\\\\\\\\\\\\end{array}\right\}$ (4.40)
      and α "sufficiently irrational"
      such as the golden mean $\alpha = \frac{\sqrt{5}-1}{2}$)
4)   Herman-ring case

While the first three cases are related to fixed (or more generally, periodic) points, the fourth case is not. It was discovered by the French mathematician M. Herman a few years ago. The Siegel-disk[11] case is named after the great German mathematician C.L. Siegel who discovered this case in 1942. His discovery gave birth to an exciting theory ($\rightarrow$ Kolmogorov–Arnold–Moser Theory) which tries, for example, to understand the stability of the solar system.

It goes beyond our possibilities and aims to describe more details here, but one more result is noteworthy: The critical points of R (i.e. solutions of $R'(z) = 0$ to be solved in $\mathbf{C} \cup \{\infty\}$) "indicate" precisely the "character" of R, i.e. which of the four cases appear (maybe all - maybe none). "Indication" means that if one of the four cases is present, then the orbit of at least one critical point z_c (i.e. $R'(z_c) = 0$) $z_c \mapsto R(z_c) \mapsto R^2(z_c)$... will "lead" to that case.

4.3.3   The quadratic family revisited

Projecting this information onto our knowledge about $f_c(z) = z^2 + c$ gathered in the first paragraph, it will become much more clearer even, what the different possibilities hidden in various choices for c are about. Firstly, note that the only critical points of the f_c are $\{0, \infty\}$. Indeed to solve $f_c'(z) = 0$ in $\mathbf{C} \cup \{\infty\}$ decomposes into two steps. The only finite solution is 0. For $z = \infty$ we must look at Eqn. (4.1) , i.e. $F_c'(u) = 0$ which for $u = 0$ corresponds to $z = \infty$. Since $f_c(\infty) = \infty$ for all c, everything about the nature of f_c will depend only

---

[11] Let $R(\alpha) = \alpha z + z^2$, then $R(0) = 0$ and $R'(0) = \alpha$. Thus, if $\alpha = \exp(2\pi i\alpha)$ and α is the golden mean, then we have an example. The reader may try to compute J_R and observe the dynamics near 0 (see [83] for details).

on the orbit

$$0 \mapsto c \mapsto c^2 + c \mapsto \dots$$

If $0 \in A_c(\infty)$, i.e. $c \notin M$, then there cannot be any of the four cases of Eqn. (4.40) , which is reflected by the fact that then $K_c = J_c$ is a Cantor set. If, however, $0 \notin A_c(\infty)$, i.e. $c \in M$, then one may have in principal one of the four cases. It is known, however, that the fourth case does not exist for f_c (all c), and it is conjectured that the first case (aside from $A_c(\infty)$) only appears if c is from the interior of M. The second and third case occur on the boundary of M. A family which is closely related to f_c is the family (sometimes called the "logistic family", see Section 3.1.2)

$$r_\lambda(z) = \lambda z(1 - z) \qquad (4.41)$$

In fact one easily derives an affine change of coordinates $h(z)$, such that

$$f_c(z) = h^{-1}(r_\lambda(h(z))).$$

This family has the advantage that $r_\lambda(0) = 0$ for all λ and $r_\lambda'(0) = \lambda$. Thus, if $\lambda = exp(2\pi i\alpha)$ and α changes between 0 and 1 , then λ will "hit" case 2 infinitely often and also case 3 infinitely often. In fact an arbitrary small perturbation of λ suffices to push situation 2 into situation 3 and vice versa, indicating that the J_c's can jump drastically when c is changed very little. In summary, if one of the three cases 1 to 3 in Eqn. (4.40) exists aside from $A_c(\infty)$ for f_c , then the orbit of 0 will lead to it and, therefore, for each choice of c only one case can appear. It may , however, happen that none of the four cases (aside from $A_c(\infty)$) appears. For example, if $c = i$, then

$$0 \mapsto i \mapsto -1 + i \mapsto -i \mapsto -1 + i \mapsto \dots$$

In other words the "critical" orbit leads to a cycle of length 2 for which one easily finds that it is repelling. That is, $A_c(\infty)$ is the only domain in the complement of J_c and in fact we know that J_c is a dendrite (see Figure 4.3e). The "fate" of critical points also explains the exotic example (4.36). One easily finds that there R has the critical points $\{0,2\}$ and

$$2 \mapsto 0 \mapsto \infty \mapsto 1 \mapsto 1, \quad \text{(and 1 is a repelling fixed point)}$$

i.e. none of the four cases is possible, which means that $J_R = \mathbf{C} \cup \{\infty\}$. Note also that the Mandelbrot set in view of critical points is just the locus of c-values for which $0 \notin A_c(\infty)$. This leads to ideas for related experiments of more general mappings:

4.3.4 Polynomials

Note that any polynomial $p(z)$ (of degree ≥ 2) has the property that $p(\infty) = \infty$ and $p'(\infty) = 0$. Thus, if we have a one-parameter family $p_c(z)$ (e.g. $p_c(z) = z^n + c$) we can identify easily a Mandelbrot-like set ($A_c(\infty)$ again denotes the basin of attraction of ∞):

$$M_{p_c} = \{c \in \mathbf{C} \ : \ \text{all finite critical points of } p_c \notin A_c(\infty)\}.$$

Actually, since ∞ is attracting, one always has that

$$J_{p_c} = \partial A_c(\infty)$$

is the Julia set of p_c. It is known (due to P. Fatou) that for any $c \in M_{p_c}$ the Julia set J_{p_c} is connected.

We mentioned earlier that any polynomial of degree 2 can be written (after an affine change of variables) as $z^2 + c$. It is easy to see that any polynomial of degree 3 (after an affine change of variables) can be written in the form

$$p_{a,b}(z) = z^3 - 3a^2 z + b, \quad \text{where} \quad a, b \in \mathbf{C}.$$

Note that $p_{a,b}$ has the critical points $\{+a, -a, \infty\}$. Let $A_{a,b}(\infty)$ be the basin of attraction of ∞ and define

$$M_{a,b} = \{(a, b) \in \mathbf{C} \ : +a \notin A_{a,b}(\infty) \text{ and } -a \notin A_{a,b}(\infty)\}.$$

$M_{a,b}$ corresponds to the Mandelbrot set of $p_{a,b}$. $M_{a,b}$ is a fractal in 4-dimensional space and plays the same role for 3rd degree polynomials as the Mandelbrot set M does for 2nd degree polynomials. For computer graphical studies one observes that

$$M_{a,b} = M_+ \cap M_-,$$

where

$$
\begin{aligned}
M_+ &= \{(a, b) \in \mathbf{C} \ : +a \notin A_{a,b}(\infty)\} \\
M_- &= \{(a, b) \in \mathbf{C} \ : -a \notin A_{a,b}(\infty)\}
\end{aligned}
$$

One can obtain fantastic images by looking at 3-dimensional cross sections of M_+ or M_-. We include one as an appetizer (see Plate 35).

4.3.5 A special map of degree four

In [83] we discuss the particular rational mapping of degree four:

$$R_c(z) = \left(\frac{z^2 + c - 1}{2z + c - 2} \right)^2$$

This mapping arises in statistical mechanical models for ferromagnetic phase transitions. One observes immediately that for all $c \in \mathbf{C}$ one has that

$$R_c(1) = 1 \quad \text{and} \quad R_c(\infty) = \infty.$$

Moreover $R_c'(1) = 0$ and $R_c'(0) = \infty$, i.e. 1 and ∞ are attracting fixed points. The critical points of R_c are $\{1, \infty, 1 - c, \pm\sqrt{1-c}, 1 - \frac{c}{2}\}$. Furthermore, $1 - \frac{c}{2} \mapsto \infty$ and $\pm\sqrt{1-c} \mapsto 0$. Thus, to study the fate of critical points it remains to investigate the orbits of $1 - c$ and 0 only. Figure 4.26 shows an experiment (and a close up), in which the critical orbits of c-values are tested. It is quite remarkable that the well known Mandelbrot set figure appears. This phenomenon is known as the universality of the Mandelbrot set (for details see [30]).

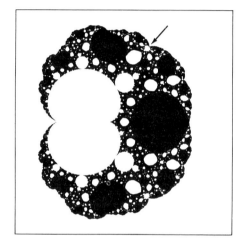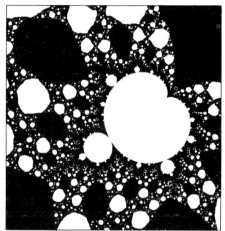

Fig. 4.26: **a.** For black values of c the orbit of 0 approaches neither 1 nor ∞ under iteration of R_c ; **b.** Blow up, for black values of c the orbit of 0 approaches 1 under iteration of R_c

4.3.6 Newton's method for real equations

Everything so far very much depends on the complex analytic structure. A natural questions is : What happens, if one considers similar problems for mappings in Euclidean space, e.g. of $\mathbf{R}^2$? Particular maps (Hénon map, piecewise linear maps) are considered in Chapter 3 . Another very interesting class of maps is obtained by Newton's method for systems of real equations, say $G(\vec{x}) = 0$, $\vec{x} \in \mathbf{R}^2$:

$$N(\vec{x}) = \vec{x} - DG^{-1}(\vec{x})G(\vec{x}).$$

Here DG is the matrix of partial derivatives at $\vec{x}$. The set analogous to a Julia set in the complex case is

$$J_N = \text{closure } \{\vec{x} : N^k(\vec{x}) \in S_G\}$$

where $S_G = \{ \vec{x} : \det(DG(\vec{x})) = 0 \}$ and det denotes the determinant of a matrix. It turns out that though J_N plays a similar role as Julia sets in the complex case do, computer-graphical studies are remarkably different. Figure 4.27 shows just two studies which document this observation. Again [83] gives concrete examples.

Fig. 4.27: Newton's method for real equations, basins of attraction of some fixed points obtained by an adaptation of LSM/J.

4.3.7 Special effects

In this concluding section we want to sketch three disjoint aspects of graphical nature

- natural and artificial homotopies

- look-up-table animations

- zoom animation without a CRAY

As our discussions of the Julia sets of $f_c(z) = z^2 + c$ has shown there is a tremendous potential for animations by just changing parameters. Also, given two maps - say f and g - a standard homotopy of the kind

$$t f(z) + (1 - t)g(z) = h_t(z), \quad 0 \leq t \leq 1 \qquad (4.42)$$

allows to transform the structures (Julia set, basins of attraction) attached to f into those attached to g.

If the image is obtained by any of the level set, binary decomposition, continuous potential or distance estimator methods, another familiar and very powerful idea is *look-up table animation*. This means that one links each level set, for example, to an entry in the look-up-table. By shifting colors through parts of the table strikingly different appearances of the same underlying structure may occur. Color map animation may then be used to continuously transform one image into the other. Naturally the structures discussed in the previous paragraph make also a nice toolkit for texture-mapping.

Our final discussion generates some ideas for zoom animations. Assume, for example, that one would like to create a zoom animation of the Mandelbrot set, say with $n(\sim 25)$ images per second. Given the amount of computing necessary for a few minutes of film a straightforward approach would be intolerable for a minicomputer (for example, a typical mini might use about 30 minutes per frame in reasonable resolution amounting to about 750 hours for 1 minute of film). Such figures seem to make supercomputing necessary. But even with a CRAY-like machine a frame could be about 30 seconds, amounting to about 12 hours of very expensive time for one minute of film. Even then the product would not be very satisfactory because near the boundary of the Mandelbrot set structures are so rich that there would be so much change from frame to frame that as a result one would see a considerable amount of noise, making some filtering absolutely necessary.

Obviously one needs a different approach. We suggest a particular interpolation method which reduces the number of images which actually have to be computed - by the time consuming level set or continuous potential methods (LSM/M , CPM/M) - dramatically. We produced a particular animation of the Mandelbrot set which zooms into a particular area of the boundary of the Mandelbrot set (seen in Figure 4.21) down to a blow-up factor of 10^8. This animation is based on only about ~ 40 (!) key frames. I. e. the actually ~ 3000 displayed frames for a 2 min animation [56] are obtained by appropriate interpolation (see Figure 4.28) which is described below.

Assume that we want to produce frames with a resolution of $R \cdot b \times b$ pixels (e.g. 680×512) for display in the animation. The idea is to compute $\nu(B)$ key frames of sufficently high resolution $R \cdot B \times B$ and interpolate the displayed frames from these (R = aspect ratio; $B > b$). The question is, how to choose B optimally? Assume that the strategy is to minimize the cost for the computation of the total number of key frames. This is legitimate if the cost for the

Keyframe k

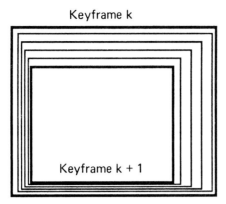

Keyframe k + 1

Fig. 4.28: Zoom interpolation from key frames.

interpolation is considerably lower. In first approximation we may then assume
that the total cost is proportional to $\nu(B)R \cdot B^2$. Given b, B and $\mu(B)$ we then
have as total zoom factor

$$\left(\frac{B}{b}\right)^{\nu(B)} = m$$

Using this relation calculus tells us that the cost is minimal for $\frac{B}{b} = \sqrt{e}$, i. e.
$B = \sqrt{e}b$. In practice, however, the cost will rather increase with increasing
magnification and as a result the optimal choice of $\frac{B}{b}$ will be somewhat smaller
than $\sqrt{e}$.

Once the key frames are obtained we suggest to compute the displayed
frames of the animation as follows: Let $K = \{1,\ldots,R \cdot B\} \times \{1,\ldots,B\}$
(respectively $D = \{1,\ldots,R \cdot b\} \times \{1,\ldots,b\}$) be a representation of the pixels
in a key frame (respectively displayed frame). Furthermore, let $w_K(i,j)$ (resp.
$w_D(i,j)$) be the complex coordinate which is represented by pixel (i,j) and let
$v_K(i,j)$ (resp. $v_D(i,j)$) be the value at pixel (i,j) (e. g. $v_K(i,j) = L(c;T)$,
where $c = w_K(i,j)$ as in Eqn. (4.26)). In first approximation one would suggest to define the values $v_D(i,j)$ by

$$v_D(i,j) = v_K(k,l)$$

where k,l are such that $| w_K(i,j) - w_D(k,l) |$ is minimal among all choices
$(k,l) \in D$

This definition, however, would produce the well known and very undesirable aliasing effects, which would appear even if all frames would have been
computed directly, i.e. not obtained by interpolation. The reason for this "noise"
is a result of the fact that small changes in w_K will produce dramatic changes in
v_K near the boundary of the Mandelbrot set and those will be inherited to v_D in
all zoom steps. To avoid this effect one has to introduce a proper "smoothing",

which is our next objective. To explain our strategy one should think of v_K as a function which is piecewise constant (on pixels).

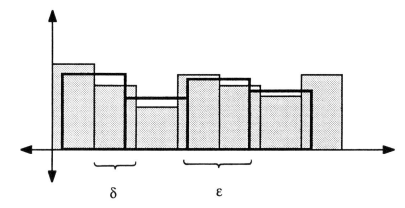

Fig. 4.29: Piecewise constant functions v_K and v_D.

Now, if δ is the size of a key frame pixel, i. e.

$$\delta = \frac{w_K(i, B) - w_K(i, 1)}{B},$$

and $\varepsilon = \sqrt{e}\delta$ is the size of a pixel in a displayed frame then Figure 4.29 illustrates the situation. Let $A_K(i, j)$ (resp. $A_D(i, j)$) denote the square domain in the complex plane of measure $\mu(A_K(i, j)) = \delta^2$ (resp. $\mu(A_D(i, j)) = e\delta^2$) which pixel (i, j) represents, see Figure 4.30.

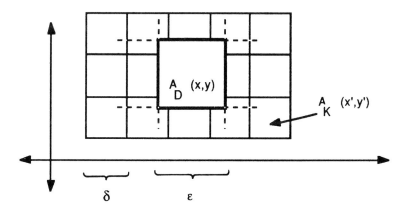

Fig. 4.30: Values of v are constant on area covered by pixels.

We then define the "smoothed" values of v_D as follows:

$$v_D(i, j) = \sum_{(k, l) \in K} \frac{\mu(A_D(i, j) \cap A_K(k, l))}{\mu(A_D(i, j))} \cdot v_K(k, l) \qquad (4.43)$$

Obviously, Eqn. (4.43) is just the integral of v_K over $A_D(i,j)$ normalized by $\mu(A_D(i,j))$. This takes care of the "noise effects". There remains one problem and that is to pass from one key frame K_n to the next one K_{n+1} without seeing a discontinuity. The homotopy in Eqn. (4.44) will solve this problem and provide appropriate key frame values:

$$v(i,j) = (1-p)v_{K_n}(i,j) + p[tv_{K_n}(i,j) + (1-t)v_{K_{n+1}}(i,j)] \quad (4.44)$$

where

$$p = \frac{1}{\mu(A_D(i,j))} \cdot \mu\left(\bigcup_{(k,l)\in K_{n+1}} (A_{K_{n+1}}(k,l) \cap A_D(i,j))\right).$$

In other words, p can be interpreted as the relative size of the intersection of pixel (i,j) with the $(n+1)$-st key frame. Finally, t should be 0 at key frame n and t should be 1 at key frame $n+1$. Combination of Eqn. (4.44) and Eqn. (4.43) solve the aliasing problem and were successfully used in [56].

Chapter 5

Fractal modelling of real world images

Michael F. Barnsley

5.1 Introduction

Mankind seems to be obsessed with straight lines. I am not sure where it all began, but I like to imagine two Druids overseeing the construction of Stonehenge, and using a piece of stretched string to check the straightness of the edges of those huge stone rectangular blocks, or the path that the light beam from the equinox sun would follow. It was at least an efficient form of quality control; for very little effort by those two Druids it kept the laborers with their eyes sharply on the job.

Whatever the real history of the matter is, it doesn't alter the fact that now a telltale and hallmark of Man's artifacts is the straight line. Look around the room at the edge for the tables, the raster scan lines of the television, and so on and on throughout all our quantitative interaction with the geometrical world which lies about us. In technology straight lines are the basic building block: to a first approximation they describe how a saw cuts, the shortest path between two points on a piece of wood. From our earliest learning moments we are encouraged to manipulate, to rotate and translate, cubes; to roll circles with toy wheels; to underline, straighten, measure along straight lines and to use graph paper. The Romans made their direct roads slice across barbarian France and England; and now our calculus is based on the use of "Functions of x" which are differentiable — when we magnify their graphs sufficiently, we assume, they all look like straight lines everywhere ! How ubiquitous they are!

The proper arena for Euclidean geometry is in the description of relative positions and of how things move. Ellipses nearly describe the paths of the planets about the sun; and lines codify the motion of galaxies away from the center of the Universe. A navigator uses vectors to put together triangles to calculate resultant velocities, and a map-maker uses laser straight lines between points on mountain peaks to help triangulate the Earth's surface.

However, the physical geometric world which actually lies about us, and was all around those two Druids, is very different from the straight lines, volumes of revolution, graphics text, etc.,with which we have trained our scientific selves to model and understand the physical world. Until the almost elusively simple observation by Mandelbrot [68] of the existence of a "Geometry of Nature" , there was no chance that we would think in a new scientific way about the edges of clouds, the profiles of the tops of forests on the horizon, the intricate moving arrangement — by no means random — of the down feathers at the joint under a bird's wing as it flies. It seems that artists made a similar transition to the one the technologists and scientists are now starting, some time ago. I quote from a book about landscape painting [21]:

> *The leaves, flowers and tendrils of Rheims and Southwell, which, in the later twelfth century, break through the frozen crust of monastic fear, have the clarity of newly created things. Their very literalness and lack of selection is a proof that they have been seen for the first time.*

Geometry is concerned with making our spatial intuitions objective, and fractal geometry extends that process. It is very hard at first, and then increasingly easy, for us to quantify, and exploit the geometrical relationships that are observed in the physical world. The theory introduced in this chapter provides us with an new ruler, which is easily passed from one of us to an other. The theory is IFS theory, and with it we can describe a cloud as clearly as can an architect describe a house. IFS theory concerns *deterministic* geometry : it is an extension of classical geometry. It uses classical geometrical entities (for example affine transformations, scalings, rotations, and congruences) to express relations between parts of generalized geometrical objects , namely "fractal subsets" of the Euclidean plane. Using only these relations IFS theory defines and conveys intricate structures.

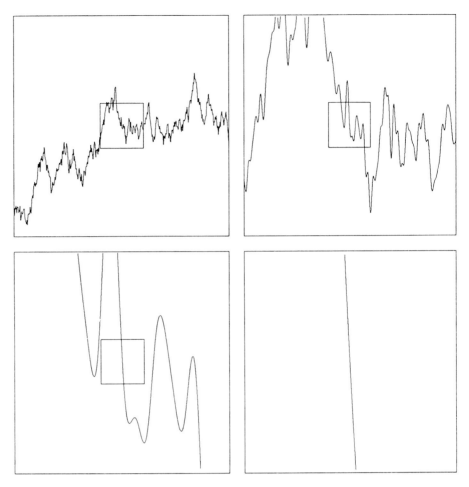

Fig. 5.1: Calculus is founded on functions whose graphs, when sufficiently magnified, look like straight lines locally.

5.2 Background references and introductory comments

The feasibility of using iterated function systems (IFS) in computer graphics was reviewed at the 1985 SIGGRAPH meeting by Demko , Naylor , and Hodges [23]. The present work concerns the development of IFS as a practical tool for the production of images including clouds and smoke, horizons, seascapes, flames , branching structures such as leaves and ferns, and simple man-made objects. Specifically, it addresses the problem of how IFS may be used to model geometrically and render two- and three-dimensional objects.

The use of fractal geometry, both deterministic and nondeterministic, to model natural scenes and objects, has been investigated by a number of authors, including Mandelbrot [68], Kawaguchi [59], Oppenheimer [80], Fournier et al. [40], Smith [96], Miller [76], and Amburn et al. [1]. The approach presented here has its roots in these investigations. However, it is characterized by a single deterministic framework which can reach a seemingly unlimited range of objects. This framework concentrates on measure theory more than it does on fractal geometry. The mathematical basis of the theory of iterated function systems occurs in a rapidly growing literature , see [5,31,48,54].

To address the proposed modeling and rendering problem, two new IFS concepts are introduced. They are (1) the Collage Theorem, which suggests an interactive geometrical modeling algorithm for finding iterated function system codes, and (2) a random iteration algorithm for computing the geometry of, and rendering, images starting from IFS codes (*RenderIFS()*).

The Collage Theorem [4] provides a means for the interactive two-dimensional modeling using IFS, and is suitable for implementation on a computer graphics workstation. The input is a two-dimensional target image, for example a polygonal approximation to the boundary of a leaf. The output from the algorithm is an IFS code which, when input to algorithm *RenderIFS()*, provides a rendition of the original target. The closeness of this rendition to the desired image depends on the efficiency with which the user is able to solve interactively a certain geometrical problem.

Algorithm *RenderIFS()* is based on an extension of the mathematical theory of IFS which permits the use of transformations which do not shrink spatial distances, as formulated in[6] and[32]. It starts from an input IFS code, explained in Section 5.3 , and with the aid of random iteration produces a deterministic geometrical object together with rendering values. Despite the use of random iteration, a unique final image is obtained once the viewing window, resolution, and a color assignment function have been specified.

The present approach contrasts with the usage of random recursive refinement algorithms to produce terrain models (see for example [40,68] and [76]), stochastic procedures used to produce clouds and textures [68] and random branching and growth models for plants ([2,59,80]): in all such cases the final product depends upon the precise random number sequence called during computation. The present algorithm *RenderIFS()* has the feature that small changes in the parameter values in the input IFS code yield only small changes in the resulting image. This is important for system independence, interactive usage, and animation. Furthermore, images vary consistently with respect to changes of viewing window and resolution. Images can be generated to a very high resolution, or equivalently viewed within a small window, without reducing to blocks of solid color.

Additional reasons for the development of IFS algorithms in computer graphics include the following: (a) IFS codes provide image compression: images that look like plants and clouds, for example, including complicated rendering, can be generated from small data bases. This means that many different images can be stored within a workstation environment. (b) IFS algorithms are such that they can be implemented in a highly parallel manner. For example, algorithm *RenderIFS()* can be run on many processors at once, all using different random number generators. This means that, in principle, natural images can be generated and manipulated in real time.

The usage of the Collage Theorem and algorithm *RenderIFS()* in the geometrical modelling and rendering of two- and three-dimensional objects is demonstrated in Section 5.6. It is shown how to produce images of clouds and smoke, chimneys and ferns.

5.3 Intuitive introduction to IFS: Chaos and measures

In Section 5.5 and 5.6 we formalize. In this section we provide informally some of the key ideas .

5.3.1 The Chaos Game : 'Heads ', 'Tails' and 'Side'

Assume that there are three points labelled 'Heads', 'Tails' and 'Side' marked arbitrarily on a piece of paper. You are the player: mark a fourth point on the paper (locate it where you like), and label it z_1. Now pretend you own a coin which, when tossed lands sometimes on its side, sometimes face up, and sometimes tails up. Toss the coin ! It comes up 'Heads'. Mark a new point z_2, half way between z_1 and the point labelled 'Heads'. Toss the coin. It comes up 'Side'. Mark a new point z_3, half way between z_2 and the point labelled 'Side'. At the nth step, toss the coin and according as to whether it comes up 'Heads', 'Tails' or 'Side', mark a new point z_{n+1} on the paper, half way between the point z_n and the point labelled with the same name as the coin shows. Keep doing this for a long time, and mark say 1,000,000 points on the paper $z_1, z_2, z_3, \ldots, z_{999,999}, z_{1,000,000}$. Now erase the first eight points $z_1, z_2, z_3, \ldots, z_8$. What do the remaining points look like? Do they make a random dust on the paper ? At first one expects nothing much !

However, what you actually see, if your pencil is sharp enough, your measuring accurate enough, and if the odds are not dreadfully against you, will be the picture in Figure 5.2.

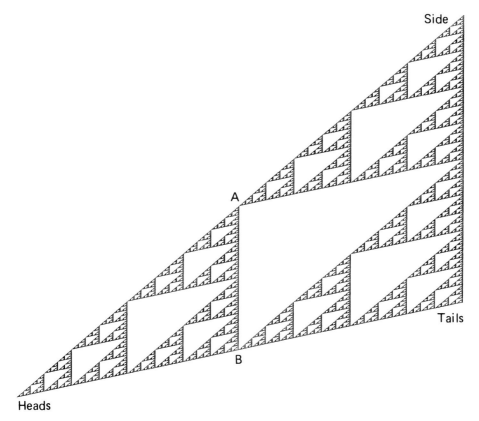

Fig. 5.2: The result of a million coin tosses when the Chaos Game is played ! A Sierpinski triangle with vertices at 'Heads', 'Tails' and and 'Side', is revealed.

Let us call the limiting picture P. Notice how P is the union of three half-sized copies of itself. In fact

$$P = w_1(P) \cup w_2(P) \cup w_3(P)$$

where, for example, w_1 is the affine transformation (for more about affine transformations see below) which takes the triangle with vertices at 'Heads', 'Tails' and 'Side', to the triangle with vertices at 'Heads', A and B respectively.

Now let us describe a generalization of the Chaos Game. Let each of w_1, w_2, $\ldots$, w_N be a transformation which takes a point z in the plane, $z \in \mathbf{R}^2$ to a new point $w(z) \in \mathbf{R}^2$. Let each transformation be such *that it moves points closer together*. That is, if $d(z_1, z_2)$ is the distance between a pair of points z_1 and z_2, then for some $0 \le k < 1$

$$d\left(w_j(z_1), w_j(z_2)\right) < k \cdot d(z_1, z_2)$$

for all pairs of points z_1 and z_2 in $\mathbf{R}^2$, for each transformation w_j. Such a transformation is called a contraction mapping . An example of such a transformation is $w : \mathbf{R}^2 \to \mathbf{R}^2$ which acts on the point $z = (x, y)$ according to

$$w(x, y) = \left(\frac{1}{2}x + \frac{1}{4}y + 1, \quad \frac{1}{4}x + \frac{1}{2}y + 0.5 \right).$$

Notice that this can also be written, in matrix notation,

$$w \begin{bmatrix} x \\ y \end{bmatrix} = \begin{bmatrix} 0.5 & 0.25 \\ 0.25 & 0.5 \end{bmatrix} \begin{bmatrix} x \\ y \end{bmatrix} + \begin{bmatrix} 1 \\ 0.5 \end{bmatrix}$$

The action of this contraction mapping on a smiling face inscribed in a triangle is shown in Figure 5.3.

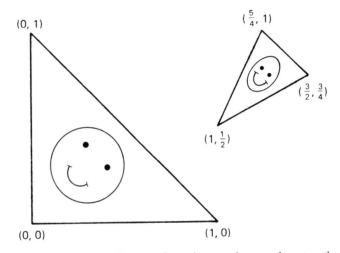

Fig. 5.3: A contractive affine transformation puts the eyes closer together.

Here is the generalization of the Chaos Game. Let $w_1, w_2, \ldots, w_N$ be contraction mappings acting on $\mathbf{R}^2$ (or a piece of paper). Choose any point z_0 and randomly pick and apply the transformations to produce a sequence of points $z_1, z_2, z_3, \ldots, z_{999,999}, z_{1,000,000}, \ldots$. For example, to obtain z_{73} we randomly choose one of the points $w_1(z_{72}), w_2(z_{72}), \ldots, w_N(z_{72})$. Then, with probability one, the sequence will converge to a unique picture, or limiting set of points. This picture, perhaps grey on a white background, will be the only one which obeys the equation (see [54])

$$P = w_1(P) \cup w_2(P) \cup \ldots \cup w_N(P) \tag{5.1}$$

Part of the result P of randomly iterating five affine transformations, which themselves were somewhat arbitrarily chosen, is shown in Figure 5.4.

Here are some points to be noticed about Figure 5.4. The structure of P is very intricate : boundaries would not be simplified by magnification they are indeed 'fractal'. P is totally fixed once one knows the transformations

Fig. 5.4: Part of the attractor, or invariant picture, made by playing the Chaos Game with five affine contraction mappings.

$w_1, w_2, \ldots, w_N$ which make it. Each affine transformation is specified by six two digit numbers — so I can tell my brother Jonathan in thirty seconds on the telephone in Australia (he knows how to play the Chaos Game), how to make exactly the same geometrical entity which we are discussing here. The image P is really complicated: it contains for example many fundamentally different holes (technically holes which, including their boundaries , cannot be mapped under diffeomorphisms from one to another). P is not self-similar: high magnification of parts of the picture will not reveal little copies of the whole picture. P does not depend upon the relative probabilities assigned to the different transformations, as long as they are all positive; (however, if any probability is too small one might have to play the Chaos Game for a very long time indeed before the whole picture P would appear).

5.3.2 How two ivy leaves lying on a sheet of paper can specify an affine transformation

It will be very helpful, for understanding how to apply the Collage Theorem and Algorithm *RenderIFS()* in Section 5.4, to see how two pictures of approximately the same flat object can be used to determine an affine transformation

$$w : \mathbf{R}^2 \to \mathbf{R}^2.$$

To illustrate the idea we start from a Xerox copy of two real ivy leaves (no copyright problems!), one big one and one small one, as shown in Figure 5.5. We wish to find the real numbers a, b, c, d, e, and f so that the transformation, in obvious notation

$$w \begin{bmatrix} x \\ y \end{bmatrix} = \begin{bmatrix} a & b \\ c & d \end{bmatrix} \begin{bmatrix} x \\ y \end{bmatrix} + \begin{bmatrix} e \\ f \end{bmatrix} = \begin{bmatrix} ax + by + e \\ cx + dy + f \end{bmatrix} \qquad (5.2)$$

has the property w(BIG LEAF) approximately equals w(LITTLE LEAF)

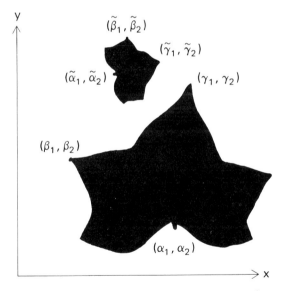

Fig. 5.5: Two ivy leaves fix an affine transformation $w : \mathbf{R}^2 \to \mathbf{R}^2$.

Begin by introducing x and y coordinate axes, as already shown in Figure 5.5. Mark three points on the big leaf (we have chosen the leaf tip, a side spike, and the point where the stem joins the leaf) and determine their coordinates (α_1, α_2), (β_1, β_2) and (γ_1, γ_2). Mark the corresponding points on the little leaf, assuming that a caterpillar hasn't eaten them, and determine their coordinates; say $(\tilde{\alpha}_1, \tilde{\alpha}_2)$, $(\tilde{\beta}_1, \tilde{\beta}_2)$ and $(\tilde{\gamma}_1, \tilde{\gamma}_2)$ respectively.

Then a, b, and e are obtained by solving the three linear equations

$$\begin{aligned}
\alpha_1 a + \alpha_2 b + e &= \tilde{\alpha}_1, \\
\beta_1 a + \beta_2 b + e &= \tilde{\beta}_1, \\
\gamma_1 a + \gamma_2 b + e &= \tilde{\gamma}_1;
\end{aligned}$$

while c, d and f sastify

$$\begin{aligned}
\alpha_1 c + \alpha_2 d + f &= \tilde{\alpha}_2, \\
\beta_1 c + \beta_2 d + f &= \tilde{\beta}_2, \\
\gamma_1 c + \gamma_2 d + f &= \tilde{\gamma}_2.
\end{aligned}$$

5.4 The computation of images from IFS codes

5.4.1 What an IFS code is

An affine transformation $w : \mathbf{R}^2 \to \mathbf{R}^2$ from two dimensional space $\mathbf{R}^2$ into itself is defined by

$$w \begin{bmatrix} x \\ y \end{bmatrix} = \begin{bmatrix} a_{11} x + a_{12} y + b_1 \\ a_{21} x + a_{22} y + b_2 \end{bmatrix} \tag{5.3}$$

where the a_{ij}'s and b_i's are real constants. If A denotes the matrix (a_{ij}), $\vec{b}$ denotes the vector $(b_1, b_2)^t$, where t denotes the transpose, and $\vec{x}$ denotes the vector $(x_1, x_2)^t$, then we write

$$w(\vec{x}) = A\vec{x} + \vec{b}$$

The affine transformation is specified by six real numbers. Given an affine transformation, one can always find a nonnegative number s so that

$$\| w(\vec{x}) - w(\vec{y}) \| \le s \cdot \| \vec{x} - \vec{y} \| \text{ for all } \vec{x} \text{ and } \vec{y}$$

We call the smallest number s so that this is true the *Lipschitz constant* for w. Here

$$\| \vec{x} \| = \sqrt{x_1^2 + x_2^2}$$

Such an affine transformation is called *contractive* if $s < 1$, and it is called a *symmetry* if

$$\| w(\vec{x}) - w(\vec{y}) \| = \| \vec{x} - \vec{y} \| \text{ for all } \vec{x} \text{ and } \vec{y}$$

It is *expansive* if its Lipschitz constant is greater than one. A two-dimensional IFS consists of a set of N affine transformations, N an integer, denoted by

$$\{w_1, w_2, w_3, \ldots, w_N\},$$

each taking $\mathbf{R}^2$ into $\mathbf{R}^2$, together with a set of probabilities

$$\{p_1, p_2, p_3, \ldots, p_N\},$$

where each $p_i > 0$ and

$$p_1 + p_2 + p_3 + \ldots + p_N = 1.$$

Let s_n denote the Lipschitz constant for w_n for each $n = 1, 2, \ldots, N$. Then we say that the IFS code obeys the *average contractivity condition* if

$$s_1^{p_1} \cdot s_2^{p_2} \cdot s_3^{p_3} \cdot \ldots \cdot s_N^{p_N} < 1.$$

An IFS code is an IFS

$$\{w_n, p_n : n = 1, 2, \ldots, N\}$$

such that the average contractivity condition is obeyed[1].

5.4.2   The underlying model associated with an IFS code

Let $\{w_n, p_n : n = 1, 2, \ldots, N\}$ be an IFS code. Then by a theorem of Barnsley and Elton [6] there is an unique associated geometrical object, denoted by $\mathcal{A}$, a subset of $\mathbf{R}^2$, called the attractor of the IFS. There is also an unique associated measure denoted by μ. This measure may be thought of as a distribution of infinitely fine sand, of total mass one, lying upon $\mathcal{A}$, as described intuitively above. The measure of a subset β of $\mathcal{A}$ is the weight of sand which lies upon β. It is denoted by $\mu(\beta)$. The underlying model associated with an IFS code consists of the attractor $\mathcal{A}$ together with the measure μ, and is symbolized by $(\mathcal{A}, \mu)$.

The structure of $\mathcal{A}$ is controlled by the affine maps $\{w_1, w_2, \ldots, w_N\}$ in the IFS code. That is, the $6 \cdot N$ numbers in the affine maps fix the geometry of the underlying model and will in turn determine the geometry of associated images. The measure μ is governed by the probabilities $\{p_1, p_2, \ldots, p_N\}$ in the IFS code. It is this measure which provides the rendering information for images.

The underlying model $(\mathcal{A}, \mu)$ may be thought of as a subset of two-dimensional space whose geometry and coloration (fixed by the measure) are defined at the finest imaginable resolution. The way in which the underlying model defines images, via projection through viewing windows onto pixels, is described in the next section. The algorithm for the computation of these images is given in Section 5.4.4.

---

[1] The notion of an IFS can easily be generalized to nonlinear mappings. See, e.g. the variants of the Inverse Iteration Method (IIM) for Julia sets in Chapter 4 and algorithm *JuliaIIM( )* in Chapter 3.

5.4.3   How images are defined from the underlying model

Let $(\mathcal{A}, \mu)$ be the underlying model associated with an IFS code. Let a viewing window be defined by

$$V = \{(x, y) : x_{min} \leq x \leq x_{max}, \quad y_{min} \leq y \leq y_{max}\}.$$

It is assumed that V has positive measure, namely $\mu(V) > 0$. Let a viewing resolution be specified by partitioning V into a grid of $L \times M$ rectangles as follows. The interval $[x_{min}, x_{max})$ is divided into L subintervals $[x_l, x_{l+1})$ for $l = 0, 1, \ldots, L - 1$, where

$$x_l = x_{min} + (x_{max} - x_{min})\frac{l}{L}.$$

Similarly $[y_{min}, y_{max})$ is divided into M subintervals $[y_m, y_{m+1})$ for $m = 0, 1, \ldots, M - 1$ where

$$y_m = y_{min} + (y_{max} - y_{min})\frac{m}{M}.$$

Let $V_{l,m}$ denote the rectangle

$$V_{l,m} = \{(x, y) : x_l \leq x < x_{l+1}, y_m \leq y < y_{m+1}\}.$$

Then the *digitized* model associated with V at resolution $L \times M$ is denoted by $\tilde{I}(V, L, M)$. It consists of all those rectangles $V_{l,m}$ such that $\mu(V_{l,m}) \neq 0$, (that is , all those rectangles upon which there resides a positive mass of sand).

The digitized model $\tilde{I}(V, L, M)$ is rendered by assigning a single RGB index to each of its rectangles $V_{l,m}$. To achieve this, one specifies a color map f which associates integer color indices with real numbers in [0,1]. Let $numcols$ be the number of different colors which are to be used. One might choose for example nine grey tones on an RGB system; then $numcols = 9$ and color index i is associated with $i \cdot 12.5\%$ Red, $i \cdot 12.5\%$ Green, and $i \cdot 12.5\%$ Blue, for $i = 0, 1, 2, \ldots, 8$. The interval [0,1] is broken up into subintervals according to

$$0 = C_0 < C_1 < C_2 < \ldots < C_{numcols} = 1.$$

Let the color map f be defined by $f(0) = 0$ and for $x > 0$ by

$$f(x) = \max\{i \; : \; x > C_i\}.$$

Let μ_{max} denote the maximum of the measure μ contained in one of the rectangles $V_{l,m}$:

$$\mu_{max} = \max_{\substack{l = 0, \ldots, L - 1 \\ m = 0, \ldots, M - 1}} \mu(V_{l,m}).$$

$\tilde{I}(V, L, M)$ is rendered by assigning color index

$$f\left(\frac{\mu(V_{l,m})}{\mu_{max}}\right) \qquad (5.4)$$

to the rectangle $V_{l,m}$.

In summary, the underlying model is converted to an image, corresponding to a viewing window V and resolution $L \times M$, by digitizing at resolution $L \times M$ the part of the attractor which lies within the viewing window. The rendering values for this digitization are determined by the measure $\mu(V_{l,m})$, (which corresponds to the masses of sand which lie upon the pixels).

5.4.4   The algorithm for computing rendered images

The following algorithm starts from an IFS code ($w_n, p_n : n = 1, 2, ..., N$) together with a specified viewing window V and resolution $L \times M$. It computes the associated IFS image, as defined in the previous section. In effect a random walk in R^2 is generated from the IFS code, and the measures $\mu(V_{l,m})$ for the pixels are obtained from the relative frequencies with which the different rectangles $V_{l,m}$ are visited. That the algorithm works is a consequence of a deep theorem by Elton [32].

An initial point (x_0, y_0) needs to be fixed. For simplicity assume that the affine transformation $w_1(\vec{x}) = A\vec{x} + \vec{b}$ is a contraction. Then, if we choose (x_0, y_0) as a fixed point of w_1 we know a priori that (x_0, y_0) belongs to the image. This is obtained by solving the linear equation

$$\begin{bmatrix} x_0 \\ y_0 \end{bmatrix} - A \begin{bmatrix} x_0 \\ y_0 \end{bmatrix} = \begin{bmatrix} b_1 \\ b_2 \end{bmatrix}.$$

An $L \times M$ array I of integers is associated with the digitized window. A total number of iterations, *num*, large compared to $L \times M$ also needs to be specified. The random walk part of the algorithm now proceeds as shown in the code of *RenderIFS()*. The $L \times M$ array I is initialized to zero. After completion of the algorithm the elements of the array I are given color index values according to Eqn. (5.4), i.e.

$$f(\frac{I[l][m]}{Imax}).$$

Providing that *num* is sufficiently large (see examples in 5.6), the ergodic theorem of Elton ensures that, with very high probability, the rendering value $I_{l,m}$ assigned to the pixel (l, m) stabilizes to the unique value defined in Section 5.4.3 as $f\left(\frac{\mu(V_{l,m})}{\mu_{max}}\right)$. It is this algorithm which is used to calculate all of the images given with this article.

ALGORITHM **RenderIFS** (x, y, P, num, xmin, xmax, ymin, ymax, I, L, M, numcols, seed)
Title	Image rendering of Iterated Function Systems

Arguments	x,y	starting point for iteration
	P[]	array of probabilities summing to 1
	num	total number of iterations
	xmin, xmax	low and high x-value of image window
	ymin, ymax	low and high y-value of image window
	I[][]	color index output array of size L by M
	L, M	image resolution in x- and y-direction
	numcols	number of colors to be used
	seed	seed value for random number generator
Variables	k, l, m, n	integer
	Imax	maximum element in I[][]
	sum	real number
Globals	Arand	rand() returns values between 0 and Arand
Functions	srand()	initialization of random numbers
	rand()	random number generator
	int()	integer part of argument
	w(x, y, k)	returns image of (x, y) under w_k

```
BEGIN
    srand (seed)
    Imax :=0
    FOR l = 0 TO L-1 DO
        FOR m = 0 TO M-1 DO
            I[l][m] := 0
        END FOR
    END FOR
    FOR n = 1 TO num DO
                                /* choose a map wk */
        r:= rand() / Arand
        sum:= P[1]
        k:=1
        WHILE (sum < r) DO
            k := k + 1
            sum := sum + P[k]
        END WHILE
                                /* apply the map wk */
        (x, y) := w (x, y, k)
        IF (x > xmin AND x < xmax AND y > ymin AND y < ymax) THEN
                                /* compute indices of pixel */
            l:= int (L * (x - xmin) / (xmax - xmin))
            m:= int (M * (y - ymin) / (ymax - ymin))
                    /* increment counters */
            I[l][m]:= I[l][m] + 1
            Imax := max (I[l][m], Imax)
        END IF
    END FOR
                                /* compute color values */
    FOR l = 0 TO L-1 DO
        FOR m = 0 TO M-1 DO
            I[l][m] := numcols * I[l][m] / (Imax+1)
        END FOR
    END FOR
END
```

Fig. 5.6: Illustrations of the Collage Theorem. The upper collage is good, so the corresponding attractor, shown upper right resembles the leaf target.

5.5 Determination of IFS codes: The Collage Theorem

In an IFS code $\{w_n, p_n : n = 1, 2, \ldots, N\}$, it is the w_n's which determine the geometry of the underlying model, i.e. the structure of the attractor, while it is the p_n's which provide rendering information through the intermediary of the measure. Here we describe an interactive two-dimensional geometric modelling algorithm for determining the w_n's corresponding to a desired model. (Some comments on the way in which the p_n's are chosen to produce desired rendering effects are made in Section 5.6.) The algorithm has its mathematical basis in the Collage Theorem [4].

The algorithm starts from a target image T which lies within a viewing window V, here taken to be $[0, 1] \times [0, 1]$. T may be either a digitized image (for example a white leaf on a black background) or a polygonal approximation (for example a polygonalized leaf boundary). T is rendered on the graphics workstation monitor. An affine transformation $w_1(\vec{x}) = A^{(1)}\vec{x} + \vec{b}^{(1)}$ is introduced, with coefficients initialized at $a_{11}^{(1)} = a_{22}^{(1)} = 0.25$, $a_{12}^{(1)} = a_{21}^{(1)} = b_2^{(1)} = 0$. The

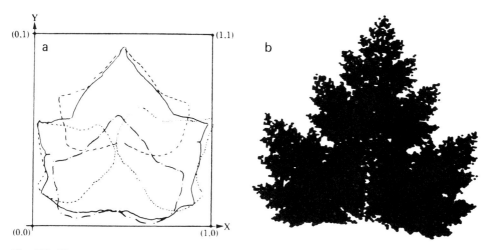

Fig. 5.7: Illustrations of how the Collage Theorem is applied. The collage in (a) determines the IFS in the Table 5.1 . In turn this fixes the attractor shown in (b).

image $w_1(T)$ is displayed on the monitor in a different color from T. $w_1(T)$ actually consists of a quarter size copy of T centered closer to the point $(0,0)$. The user now interactively adjusts the $a_{i,j}^{(1)}$'s using keypresses or a mouse, so that the image $w_1(T)$ is variously translated, rotated, and sheared on the screen. The goal of the user is to transform $w_1(T)$ so that it lies over part of T : that is, $w_1(T)$ is rendered as a subset of the pixels which represent T. It is important that the dimensions of $w_1(T)$ are smaller than those of T, to ensure that w_1 is a contraction. Once $w_1(T)$ is suitably positioned it is fixed, a new subcopy of the target $w_2(T)$ is introduced. Then w_2 is interactively adjusted until $w_2(T)$ covers a subset of those pixels in T which are not in $w_1(T)$. Overlap between $w_1(T)$ and $w_2(T)$ is allowed, but for efficiency it should be as small as possible.

$w_i(z) = s_i z + (1 - s_i)\, a_i$	
$s_1 = 0.6$	$a_1 = 0.45 + 0.9\,i$
$s_2 = 0.6$	$a_2 = 0.45 + 0.3\,i$
$s_3 = 0.4 - 0.3\,i$	$a_3 = 0.60 + 0.3\,i$
$s_4 = 0.4 + 0.3\,i$	$a_4 = 0.30 + 0.3\,i$

Table 5.1: IFS parameters for Figure 5.7 .

In this way the user determines a set of contractive affine transformations $\{w_1, w_2, \ldots, w_N\}$ with this property : the original target T and the set

$$\tilde{T} = \bigcup_{n=1}^{N} w_n(T)$$

are visually close, while N is as small as possible. The mathematical indicator of the closeness of T and $\tilde{T}$ is the Hausdorff distance $h(T,\tilde{T})$ defined below; and by "visually close" we really mean "$h(T,\tilde{T})$ is small". The maps $\{w_1, w_2, \ldots, w_N\}$ thus determined are stored. The Collage Theorem, stated below, then assures us that the attractor of any IFS code, which uses these maps, will also be "visually close" to T. Moreover, if $T = \tilde{T}$, more precisely if $h(T,\tilde{T}) = 0$, then $\mathcal{A} = T$. Thus, the algorithm provides IFS codes such that the geometry of the underlying model resembles that of the target.

The algorithm is illustrated in Figure 5.6 which shows a polygonal leaf target T at the upper and lower left. In each case T has been approximately covered by four affine transformations of itself. The task has been poorly carried out in the lower image and well done in the upper image. The corresponding attractors are shown on the left hand side: the upper one is much closer to the geometry of the target because the collage is better. The process is also illustrated in Figure 5.7 which shows (a) a collage, (b) the IFS which it determines (in complex notation), and (c) the attractor of the IFS.

The precise statements which govern the above algorithm are as follows. The *Hausdorff distance* $h(A, B)$ between two closed bounded subsets A and B of $\mathbf{R}^2$ is defined as

$$h(A, B) = \max \left\{ \max_{\vec{x} \in A} \min_{\vec{y} \in B} \| \vec{x} - \vec{y} \|; \max_{\vec{y} \in B} \min_{\vec{x} \in A} \| \vec{y} - \vec{x} \| \right\} \qquad (5.5)$$

The Collage Theorem makes a statement about the Hausdorff distance between the attractor $\mathcal{A}$ of an iterated function system and the target set T.

Collage Theorem *[4]* : *Let* $\{w_n, p_n : n = 1, 2, \ldots, N\}$ *be an IFS code of contractive affine maps. Let* $s < 1$ *denote the largest Lipschitz constant for the maps. Let* $\varepsilon > 0$ *be any positive number. Let* T *be a given closed bounded subset of* $\mathbf{R}^2$ *, and suppose the maps* w_n *have been chosen so that*

$$h\left(T, \bigcup_{n=1}^{N} w_n(T)\right) < \varepsilon.$$

Then

$$h(T, \mathcal{A}) < \frac{\varepsilon}{1 - s}$$

where $\mathcal{A}$ *denotes the attractor of the IFS.*

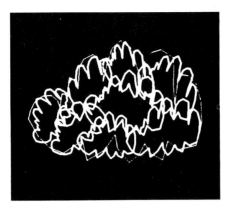

Fig. 5.8: Illustration of the Collage Theorem. Eight affine transformations of a polygonalized cloud boundary have been overlayed on the target. Images rendered from the IFS code thus determined are shown in Figures 5.9 and 5.10 .

Fig. 5.9: Cloud image rendered with algorithm *RenderIFS()* from an IFS code determined in Figure 5.8. One transformation has twice the probability of each of the others and produces an apparent illumination direction for the image.

Fig. 5.10: The same as Figure 5.9, but a different cloudlet has been given increased probability, changing the apparent light source.

Fig. 5.11: A cloud and two successive zooms (Figures 5.12,5.13) from an IFS code determined in Figure 5.8. This illustrates how the same small database provides a model which can be viewed at many resolutions.

Fig. 5.12: Zoomed view of part of Figure 5.11.

Fig. 5.13: Zoomed view of part of Figure 5.12.

5.6   Demonstrations

5.6.1   Clouds

Figure 5.8 shows eight affine transformations of a polygonalized cloud boundary. Using algorithm *RenderIFS()* they have been overlayed on the original target and determine a two-dimensional IFS code $\{w_n, p_n : n = 1, 2, \ldots, 8\}$. The probabilities are chosen proportional to the areas of the corresponding cloudlets, with the exception of one which is given twice this value. A color map $f(x)$ which assigns a grey-scale, from black to white, to the interval $0 \leq x \leq 1$ is used. Two corresponding images, rendered using the algorithm *RenderIFS()* with $num = 2 \cdot 10^6$, are shown in Figures 5.9 and 5.10. In each case a different cloudlet has been given increased probability and results in the simulation of a cloud illuminated from the direction of this cloudlet.

The ability of an IFS model to provide visually meaningful images on several different scales is illustrated in Figures 5.11, 5.12, and 5.13, which show successive zooms corresponding to the same IFS code. The underlying database, the IFS code, can be expressed in less than 100 bytes, other than viewing parameters. The number num of iterations is increased with magnification to keep the number of points landing within the viewing window constant.

5.6.2   Landscape with chimneys and smoke

Plates 33 and 34 show two views computed from the same underlying model. The second view is a close-up on the smoke coming out of one of the chimneys in the first view. The two-dimensional IFS codes consist of : one cloud (8 affine transformations), one background including the horizon (10 affine transformations), three different smoke plumes (5 transformations each) , three different chimneys (6 affine transformations each), and one field. The probabilities were chosen proportional to the determinants of the transformations, i.e. $|a_{11}a_{22} - a_{12}a_{21}|$ if w is given as in Eqn. (5.3) . The images are overlayed from back to front. The whole IFS database including color assignment functions, can be succinctly expressed using less than 2000 bytes, which means that, at the resolution used in each figure, the compression ratio is better than 500:1, whereas a standard run length encoder applied to an individual image here yields no better than 5:1.

The geometric modelling required to find the IFS codes used here was achieved using algorithm described in Section 5.5 applied to polygonalized target images. The latter were derived from tracings of components of pictures in an issue of National Geographic. The number num of iterations for Plate 33 was $4 \cdot 10^6$ and for the other plates it was maintained proportional to window area.

Fig. 5.14: The image at bottom left is computed using sixteen two dimensional affine maps but is more complicated than it first appears. A blow-up of a single branch is shown at bottom right — notice how it is not made of copies of the image of which it is part. Further successive blow-ups are shown at upper-left and upper-right respectively. IFS encoding allows for nearly independent geometric modelling on several scales in the same image.

5.6.3 Vegetation

The algorithms of this chapter may be applied to model and render a wide range of plants and leaves, as illustrated in Figures 5.14 to 5.16. Figure 5.14 shows a fern together with several close-ups, and demonstrates that an IFS encoded image need not be self-similar. Moreover, there is no simple plant growth model for this image. This contrasts with the leaf and associated vein structure shown in Figure 5.15. These were obtained by application of the geometric modelling algorithm to a leaf given by Oppenheimer [80]. The IFS code for the veins consists of three affine transformations, while the filled-in leaf, which could have been obtained by "painting" the region defined by the veins, is the attractor for six affines. This illustrates that some deterministic branch and growth models can be reformulated in terms of IFS theory: in such cases new algorithms, such as *RenderIFS()*, become available for rendering.

Fig. 5.15: Demonstrates how branch and growth models are subsumed by IFS theory. The IFS codes for these images were obtained by applying the geometrical modelling algorithm to the leaf image given by Oppenheimer [80].

Plate 32 shows four views of the same three-dimensional fern. The IFS code consists of four three-dimensional affine transformations which were obtained by modifying output from the geometric modelling algorithm. Views are computed directly from the IFS code using algorithm *RenderIFS()*, as illustrated in Figure 5.17. The IFS used for the underlying two-dimensional model consists of four affine maps in the form

$$w_i \begin{bmatrix} x \\ y \end{bmatrix} = \begin{bmatrix} r\cos\theta & -s\sin\psi \\ r\sin\theta & s\cos\psi \end{bmatrix} \begin{bmatrix} x \\ y \end{bmatrix} + \begin{bmatrix} h \\ k \end{bmatrix} (i = 1, 2, 3, 4),$$

see Table 5.2, and also Figure 5.17.

On a 68020 based workstation a single view of the three dimensional fern requires one minute of computation. Figure 5.16 shows an IFS encoded maple

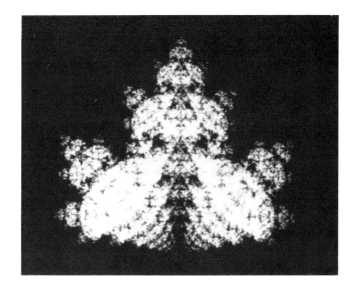

Fig. 5.16: This image illustrates how changing the probabilities on an IFS encoded maple leaf (four affine maps) can alter the rendering without changing the geometry.

Map	Translations		Rotations		Scalings	
	h	k	θ	ψ	r	s
1	0.0	0.0	0	0	0.0	0.16
2	0.0	1.6	-2.5	-2.5	0.85	0.85
3	0.0	1.6	49	49	0.3	0.3
4	0.0	0.44	120	-50	0.3	0.37

Table 5.2: Fern IFS.

leaf. Different choices of probability lead to different renderings without changing the geometry.

Fig. 5.17: The Black Spleenwort fern. One of the four affine transformations in the IFS which generates the set is illustrated. The Collage Theorem provides the other three. Observe that the stem from C to $\tilde{C}$ is the image under one of the affine transformations of the whole set. The original IFS used here was found by Michael Barnsley.

Appendix A

Fractal landscapes without creases and with rivers

Benoit B.Mandelbrot

This text tackles three related but somewhat independent issues. The most basic defect of past fractal forgeries of landscape is that every one of them fails to include river networks. This is one reason why these forgeries look best when viewed from a low angle above the horizon, and are worst when examined from the zenith. This is also why achieving a *fully random* combined model of rivers and of mountains will be a major advance for computer graphics (and also perhaps for the science of geomorphology). The main goal of this text is to sketch two *partly random* solutions that I have developed very recently and Ken Musgrave has rendered.

A second very different irritating defect is specific to the method of landscapes design (called midpoint displacement or recursive subdivision) which is described in [40]. It has gained wide exposure in "Star Trek II," and is fast, compact, and easy to implement. Its main shortcoming is that it generates "features" that catch the eye and, by their length and straightness, make the rendered scene look "unnatural." This defect, which has come to be called "the creasing problem," is illustrated in the body of the present book. It was first pointed out in [69], and its causes are deep, so that it *cannot* be eliminated by cosmetic treatment. This text sketches several methods that avoid creasing from the outset. (We were not aware of the fact that [76] had addressed this problem before us. His method is somewhat similar to ours, but there is no face-to-face overlap.)

A third defect present in most forgeries is that the valleys and the mountains are symmetric because of the symmetry of the Gaussian distribution. Voss's pictures in [69] have shown that this undesirably condition is greatly improved by after-the-fact non linear processing. Furthermore, our "mountains with rivers" involve extreme asymmetry, meaning that this paper tackles this issue head on. But, in addition, this text sketches a different and much simpler path towards the same end. It shows that it suffices to use the familiar midpoint displacement method with displacements whose distribution is itself far from being Gaussian, in fact are extremely asymmetric.

In the search for algorithms for landscape design, hence in this text, a central issue is that of *context dependence.* The familiar midpoint displacements method is context *independent,* which is the source both of its most striking virtue (simplicity) and of its most striking defect (creasing). In real landscapes, however, the very fact that rivers flow down introduces very marked context dependence; it is theoretically infinite, and practically it extends to the overall scale of the rendering.

This introduction has listed our goals in the order of decreasing importance, but this is also the order of increasing complexity, and of increasing divergence from what is already familiar to computer graphics experts. Therefore, the discussion that follows is made easier by moving towards increasing complexity. It is hoped that the new procedures described in this text will inject fresh vitality into the seemingly stabilized topic of fractal landscape design. In any event, landscape design deserves investigation well beyond what is already in the record, and beyond the relatively casual remarks in this paper.

A.1 Non-Gaussian and non-random variants of midpoint displacement

A.1.1 Midpoint displacement constructions for the paraboloids

One of the purposes of this text is to explain midpoint displacement better, by placing it in its historical context. The Archimedes construction for the parabola, as explained in the Foreword, can be expanded to the surface S closest to it, which is the isotropic paraboloid $z = P(x, y) = a - bx^2 - by^2$. Here, the simplest initiator is a triangle instead of the interval, the section of S along each side of the triangle being a parabola. A second stage is to systematically interpolate at the nodes of an increasingly tight lattice made of equilateral triangles. This procedure, not known to Archimedes, is a retrofit to non-random

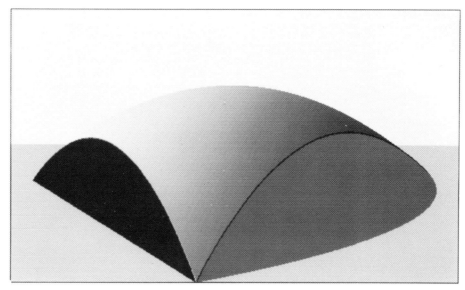

Fig. A.1: Mount Archimedes, constructed by positive midpoint displacements.

Fig. A.2: Archimedes Saddle, constructed by midpoint displacements.

classical geometry of the method Fournier et al. use to generate fractal surfaces. When all the interpolations along the cross lines go up, because $b > 0$ (resp., down, because $b < 0$), we converge to *Mount Archimedes* (Figure A.1) (resp., to the *Archimedes Cup*, which is the *Mount* turned upside down.) Suitable slight changes in this construction yield the other paraboloids. In reduced coordinates, a paraboloid's equation is $z = P(x, y) = a - bx^2 - cy^2$, with $b > 0$ in all cases, and with $c > 0$ if the paraboloid is elliptic (a cap), but $c < 0$ if it is hyperbolic (a saddle). By suitable "tuning" of the up and down midpoint displacements, one can achieve either result. The *Archimedes saddle* is shown on Figure A.2.

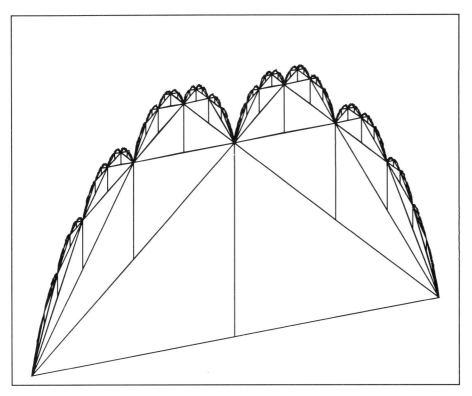

Fig. A.3: Takagi profile, constructed by positive midpoint displacements.

A.1.2   Midpoint displacement and systematic fractals: The Takagi fractal curve, its kin, and the related surfaces

Long after Archimedes, around 1900, the number theorist Teiji Takagi resurrected the midpoint displacement method, but with a different rule for the relative size (divided by δ) of the vertical displacements: Instead of their following the sequence 1 (once), 4^{-1} (twice), 4^{-2} (2^2 times) ... 4^{-k} (2^k times), etc., he made the relative sizes be 1 (once), 2^{-1} (twice), 2^{-2} (2^2 times) ... 2^{-k} (2^k times), etc.. The effect is drastic, since the smooth parabola is then replaced by the very unsmooth Figure A.3, which is an early example of a function that is continuous but non differentiable anywhere. Then a mathematician named Landsberg picked w arbitrarily in the interval $1/2 < w < 1$ and performed midpoint displacements of 1 (once), w (twice), ... w^k (2^k times), etc. ... Figure A.4 corresponds to $w = .6$.

These curves have the fractal dimension $2 - |\log_2 w|$. As w ranges from $1/2$ to 1, this dimension ranges from 1 (a borderline curve of dimension 1, but of logarithmically infinite length) to 2.

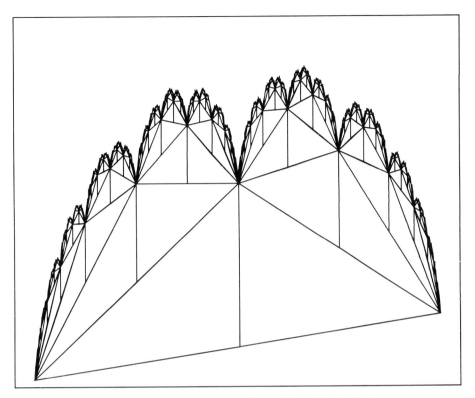

Fig. A.4: Landsberg profile, constructed by positive midpoint displacements.

(When $0 < w < 1/2$ but $w \neq 1/4$, the curves are rectifiable, with a fractal dimension saturated at the value 1. Left and right derivatives exist at all points. The two derivatives differ when the abscissa is dyadic, that is, of the form $p2^{-k}$, with an integer p, but there is a well defined derivative when the abscissa is non-dyadic.)

Again, let us retrofit the Takagi curve à la Fournier et al.. Its surface counterpart, which we call *Mount Takagi,* is shown as Figure A.5. A remarkable fact is revealed when we examine this surface from a low enough angle, relative to the vertical scale in the picture, and from a horizontal angle that avoids the lattice directions. At ridiculously low intellectual cost, we have delivered a realistic first-order illusion of a mountain terrain (other viewers may be more strongly taken by botanic illusions). However, seen from a high vertical angle, Mount Takagi reveals a pattern of valleys and of "cups" without outlets, which is totally unrealistic. From a horizontal angle along the lattice directions, the illusion also collapses altogether: the original triangular grid used for simulation overwhelms our perception. Inspecting Mount Takagi from a randomly chosen low angle yields a surprising variety of terrains that do not immediately seem to be all the same.

Fig. A.5: Mount Takagi, constructed by positive midpoint displacements.

The lesson we just learned happens to be of much wider consequence. Every one of the fractal landscapes explored thus far is best when viewed at low angle.

(The question of what is a high or a low angle, compared to the vertical, is related to the counterpart of the quantity δ in the definition of Mount Archimedes, hence to the fact that our various reliefs are self-affine, rather than self-similar.)

A.1.3 Random midpoint displacements with a sharply non-Gaussian displacements' distribution

In random midpoint displacement (Fournier et al.), the structure of valleys and ridges is less accentuated than it is with Mount Takagi, yet it remains apparent and bothersome. It can be attenuated by modifying the probability distribution of the random midpoint displacements. The Gaussian distribution is symmetric, hence a fractional Brownian surface can be flipped upside down without changing statistically. Actual valley bottoms and mountain crests *are not* symmetric. This difficulty had led Voss to introduce non linear after the fact transformations. The resulting non Gaussian landscapes (Plates 8 and 10) are stunning, but contrived. If one abandons the straight and true fractional Brown ideal, one may as well replace the Gaussian by something else.

The first suitable non symmetric distribution we have played with is the asymmetric binomial of parameter p: $X = 1$ with the probability p and

Fig. A.6: First fractal landscapes constructed by random midpoint displacements with a very skew distribution.

Fig. A.7: Second fractal landscapes constructed by random midpoint displacements with a very skew distribution.

$X = -p/(1 - p)$ with the probability $1 - p$. Values $p < 1/2$ yield flat valleys and peaky mountains, and values $p > 1/2$ yield flat mountains and deep valleys. Replacing p by $1 - p$ flips a relief upside down.

A second possibility that is easy to use and with which we have played has been the "reduced exponential -1", and a third has been the Gamma distribution with an integer parameter b, that is the "sum of b reduced exponentials $-b$".

These very simple changes beyond the Gaussian have proved to lead to a very marked increase in realism. Examples due to Réjean Gagné are shown on Figures A.6 and A.7.

A.2   Random landscapes without creases

A.2.1   A classification of subdivision schemes: One may displace the midpoints of either frame wires or of tiles

Irrespective of the issue of distribution, a second development begins by carefully examining the different published implementations of random midpoint displacement, to see what they have in common and how they differ.

First common feature. Starting with an initial lattice of nodes or vertices, one "nests" it within a finer lattice, which contains both the "old vertices" where $H(P)$, the height at P is already known, and many "new vertices" where $H(P)$ must be interpolated.

Second common feature. In order to compute $H(P)$ at N points of a midpoint fractal requires a number of calculations proportional to N. By contrast, the best-controlled approximations to fractional Brownian motion require $N \log N$ calculations. The factor $\log N$ in the cost of computing seems to us of low significance compared to the cost of rendering. But this seems, somehow, to be a controversial opinion.

A distinguishing feature. Various methods of interpolation fall into two kinds I propose to call *wire frame* displacement and *tile* displacement. (Some physicists among the readers may be helped by knowing that the distinction echoes that between *bond* and *site* percolation).

In *wire frame midpoint displacement,* the surface is thought of as a "wire frame", a collection of intervals, and one displaces the midpoint of every interval. The procedure is known to be context independent.

In *tile midpoint displacement,* the surface is thought of as a collection of tiles, and one displaces points in the "middle" of every tile. The procedure will be seen to context dependent.

The only possible implementation of pure frame midpoint displacement is the approach of Fournier et al., as applied to the subdivision of triangles.

When the initial lattice is made of squares, it is necessary to displace the relief, not only at the frame midpoints, but also at the square centers, which are midpoints of tiles. The resulting construction is a frame-tile hybrid.

In the body of this book, the desire for a cleaner algorithm to interpolate within a square lattice has pushed the authors away from the hybrid frame-tile approach, and has led them to a tile midpoint procedure. (I had failed to notice they had done so, until told by Voss.)

A.2.2   Context independence and the "creased" texture

The major difference between frame and tile midpoints involves *context dependence*, a notion from formal linguistics that has been brought into graphics in [96].

Frame midpoint fractals are context independent, in the sense that the only external inputs that affect the shape of a surface within a triangle are the altitudes at the vertices. Everyone in computer graphics knows this is an extraordinary convenience.

On the other hand, the fractional Brown surfaces incorporate an infinite span of context dependence, so that interpolating them within even a small area requires some knowledge over an unboundedly wide surrounding area. Relief models that include long range river networks, necessarily share this property. It is essential to landscape modelling.

What about tile midpoint fractals? It is important to know that they are context dependent.

Unfortunately, frame midpoint fractals have a most unwelcome texture. It reminded me of folded and crumpled paper, and (after first arguing back) Fournier et al. have acknowledged my criticism. Fractional Brown surfaces *do not* have this undesirable "creasing" texture. This discrepancy has led me long ago to wonder whether or not context dependence is the sole reason for creasing. Now we have a counter example: the implementations of tile midpoint fractals in this book are context independent, yet they possess a "creasing" texture, though perhaps to a less extreme degree.

Digging deeper, the undesirable texture could perhaps be traced to either of the following causes. A) "The Course of the Squares": The programmers' and the physicists' "folklore" suggests that spurious effects due to the lattice structure tend to be strongest in the case of square lattices in which privileged directions number two. This suggests trying triangular or hexagonal lattices, in which privileged directions number three. B) "The Nesting of the Frames". The strongest texture problems always appear along the frames that bound the largest tiles. Let us dwell on this point. As has been previously observed, the *tile vertices* at the k-th stage of construction are always nested among the tile vertices at all later stages. But, in addition, triangular tiles have the far stronger property that the network of *tile sides* at the k-th stage of construction is nested in (i.e., part of) the network of tile sides at the $(k + 2)$-th stage, the $(k + 4)$-th, etc.. I believe this feature is responsible for creasing and propose to demonstrate how it can be avoided.

My own recent renewal of active interest in fractal reliefs happens to provide tests of both of the above explanations of creasing. A first approach avoids squares and uses the classical triangular tiles, whose frames *are* nested. A second approach uses tiles that are based on the hexagon, but are actually bounded by fractal curves; their frames *are not* nested.

A.2.3   A new algorithm using triangular tile midpoint displacement

With triangles, the aim is to construct a fractal surface whose heights at the vertices of an initial triangular lattice are independent identically distributed random variables, while its values within each tile are interpolated at random. The result is said by physicists to have a "cross-over scale" that is precisely equal to the initial triangles' side: above this scale, the surface is statistically stationary, and below this scale it is more like fractional Brownian motion.

In the first stage, Figure A.8a, the height at the center of the triangle ABC in the first stage lattice is given the value

$$\frac{1}{3}[\,H(A) + H(B) + H(C)\,] + \text{a random midpoint displacement.}$$

This determines $H(P)$ at the vertices of a second order triangular lattice, which is turned by $30°$ and whose sides are smaller in the ratio $1/\sqrt{3}$. The second construction stage, Figure A.8b, takes averages in every tile and adds a suitable midpoint displacement. Note that two values will have been interpolated along each side of ABC, and that they will depend on initial values beyond the triangle ABC. The third stage lattice, Figure A.8c is again triangular and rotated by $30°$, hence it nests the first stage lattice. The third construction stage within ABC is also affected by initial values beyond this triangle.

Conceptually, one must also combine interpolation with extrapolation. The latter amounts to changing the character of the values of $H(P)$ on the nodes

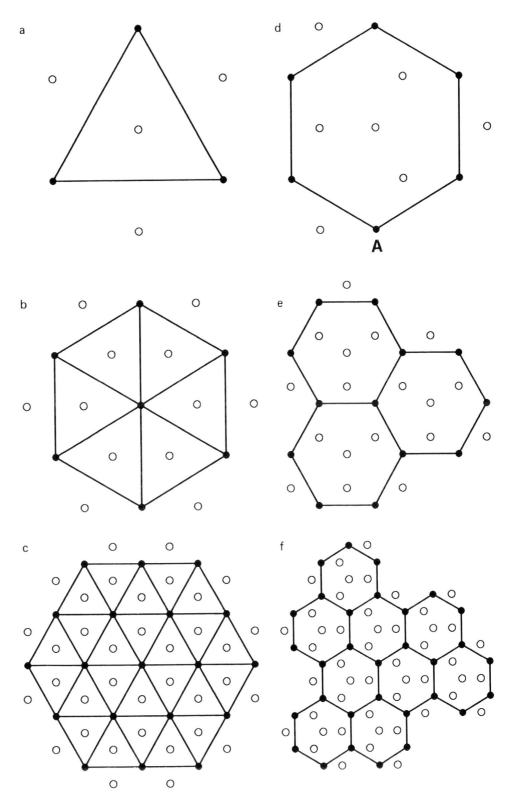

Fig. A.8: Triangular (left) versus hexagonal (right) tiles and midpoint displacements.

of the initial lattice. Instead of viewing them as stationary, extrapolation views them as having been interpolated, during a previous negative-numbered stage, from the values on the vertices of a looser lattice. If extrapolation continues indefinitely, the result is a surface in which $H(P')$ and $H(P'')$ are slightly but not negligibly correlated even when the points P' and P'' are very far from each other. This is precisely what holds for fractional Brownian motion.

A.2.4   A new algorithm using hexagonal tile midpoint displacement

The main novelty is that tiles are chosen so that their frames do *not* nest. The reason lies in the following familiar fact. While a square is subdivided into four squares and a triangle into nine triangles after two stages, a hexagon *cannot* be subdivided exactly into smaller lattice/hexagonal hexagons. However, an approximate subdivision can be done by using tiles that are near-hexagonal, but whose boundaries are fractal curves. The simplest example is illustrated by Plate 73 of [68] where it is shown being drained by an especially simple river tree that we shall encounter again in Section A.3. This tile divides into 3 subtiles of smaller size but of identical shape. (Plate 71 of [68] shows a different such tile, divisible into sevens).

Starting with a regular hexagon, the first stage of construction of this tile is shown by comparing the bold lines on Figures A.8d and A.8e: here the side that one meets first (when moving from A clockwise) is rotated by $30°$; it may have been rotated by either $+30°$ or $-30°$ and the remainder of this construction stage is fully determined by this single choice. At the same time, Figure A.8d, $H(P)$ is interpolated at the center of the initial hexagon by the equal weights average of six neighbors displacement. But $H(P)$ is also interpolated by suitable averages at three other points off-center, where three sub-hexagons meet (hence *"mid-point"* is not synonymous with *"center"*.) The averaging weights can be chosen in many different ways, with considerable effect on the texture. Displacements are added to all these weighted averages. The next interpolation stage by comparing the bold lines, shown on Figures A.8e and A.8f, again involves an arbitrary decision between $+30°$ and $-30°$, thus the rotation is $-30°$. At each stage, the boundary of the modified hexagon becomes increasingly crumpled, and it never nests in the combination of higher stage boundaries. While an undesirable texture may well be present along these crumpled curves, the fact will matter little, because these curves fail to catch the eye, contrary to straight lines that stand out.

A.3 Random landscape built on prescribed river networks

The idea behind the two landscape algorithms to be described now is to first construct a map, with its river and watershed trees, and then to "fill in" a random relief that fits the map. The two implementations available today have the weakness that the maps themselves fail to be random.

A major ingredient of this approach (beyond the current implementation) amplifies on the already discussed asymmetry between valleys and mountains, by drawing very asymmetric rivers and watersheds. The relief along a watershed can and should go up and go down, with local minima at the sources, and this requirement can be (and will be in this text) implemented by a randomized procedure of positive midpoint displacements. On the other hand, the relief along a river necessarily goes downstream, hence must follow a fractal curve that differs profoundly in its structure from anything that has been used so far in this context.

A.3.1 Building on a non-random map made of straight rivers and watersheds, with square drainage basins

Our first example of prescribed river and watershed trees is very crude, but worth describing, because it is easy to follow. It is suggested by Figure A.9, which reproduces Plate 65 of [68], rotated by 45° clockwise. In each quarter of it one sees thin black triangles: one is of length 1 and runs along a diagonal, two are of length 2^{-1}, eight of length 2^{-2}, etc. ... Each triangle is meant to represent a straight river flowing towards its very short side. Similarly, one sees thin white triangles, meant to represent a straight watershed. One builds up a non-random highly regular map by making all those triangles become infinitely thin.

Relief along the watershed. This is the most conspicuous feature in the rendering of the previous relief models. This is also where the midpoint displacement curves are least controversial, and we may as well use them to model relief here. The local minima along this relief are sharp cusps, they will be the rivers' sources, and the map prescribes their positions. The simplest implementation uses positive displacements, and consists in modeling the relief along the watersheds by the Landsberg functions defined in Section I. One may also multiply the positive midpoint displacement by a random factor that is positive, hence does not displace the sources.

Fig. A.9: Straight rivers and watersheds and square drainage basins.

Relief along rivers. This is where we must bring in something new, namely fractal curves that never go up. The simplest implementation is called binomial non-random of proportion $p < 1/2$. Here, Figure A.10a a river's altitude loss between its source and its mouth is apportioned in the ratio $1 - p$ in the upstream half, and $p < 1 - p$ in the downstream half. Similar apportioning, Figure A.10b on, is carried out along every segment of a river between points whose distances from the source are of the form $q2^{-k}$ and $(q+1)2^{-k}$. This yields for each river profile a scalloped line that is illustrated on the top line of Figure A.10. Each river is flat near its mouth and vertical near its spring, a feature that exaggerates on everyone's perception of reality, but does not do it massive injustice.

Needless to say, the fixed p may be replaced by a variable randomly selected multiplier.

Fractal structure of the saddle points. These points are exemplified by the point $(1/8, 1/8)$ on Fig. A.9, after the figure has been mentally transformed

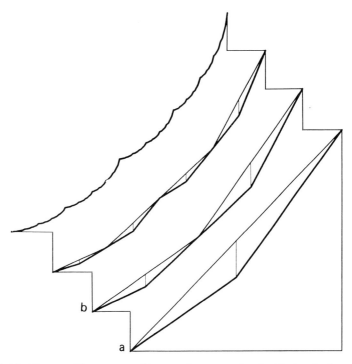

Fig. A.10: River profile constructed by negative proportional midpoint displacements.

so all the triangles are infinitely thin. Along each of the two watersheds (which become lines of slopes 45° or −45° on the map), there is a local minimum at this saddle point. The four rivers are orthogonal to the watersheds (have slopes 0 or 90° on this map), and on them the saddle point is a source and an absolute maximum.

Fractal structure of the points where rivers meet. Three rivers that meet flatten just upstream of their point of confluence and the river that emerges has a steep "waterfall" just downstream of this point. Each watershed reaches a minimum.

The initiator of the construction. The initial altitudes at the four vertices of the square must be selected to satisfy the following constraints: the mouth has the lowest altitude (think of it as sea level), the diagonally opposed point has some altitude $h > 0$, and the two remaining points have altitudes that exceed ph, but are otherwise arbitrary. Join each vertex to its neighbors, not only on the map but in relief, and join the mouth to the opposite vertex, again in relief. The resulting initial prefractal relief is a single drainage basin which projects upon two straight isosceles triangles joined by their hypotenuses.

The first stage of construction. The first step of this stage is to break each of the triangles of the initiator into two, by moving the altitude above the midpoint of the square down, from the initiator's altitude $h/2$ to the lower value ph. The effect of this step is to draw two rivers that are orthogonal to the original one, and slope towards the center. The next step is to move the midpoints of the sides of the initial squares up by a quantity w. The effect is to split each straight watershed interval into two intervals, and to add four additional watershed intervals, each going from one of the four new sources to the midpoint of the initial river. As a result, each triangle in the initiator will have split into four. The outcome of this first stage of construction consists in four squares on the map, and on each square a drainage basin that has the properties we have required of the initiator relief.

The second stage of construction. It can proceed exactly as the first. The final relief is defined by the two parameters p and w. Needless to say, again, everything can be randomized.

A.3.2   Building on the non-random map shown on the top of Plate 73 of "The Fractal Geometry of Nature"

This map in [68] was already referenced in this Appendix, when describing the second construction of a creaseless fractal forgery without rivers. Now we construct rivers and water sheds simultaneously, as on Figure A.11. The map of the initiator, Figure A.11a, is a hexagonal drainage basin, crossed by a river that joins the mouth S_0 to the source S_2. One supposes that the altitudes $H(S_0)$, $H(S_2) > H(S_0)$ and $H(S_4) > H(S_2)$ are known.

The first stage of construction. First, the altitudes at the midpoints S_1, S_3 and S_5 along the watershed are interpolated by positive midpoint displacement (either the Landsberg procedure, or a randomized version of it). This yields $H(S_1)$, $H(S_3)$ and $H(S_5)$. Next, the originally straight river is replaced by a Y-shaped river, as marked. Figure A.11b, the altitude at the midpoint of the hexagon is determined in the same spirit as in the case of square drainage basins: by dividing the difference or altitude between the original endpoints S_0 and S_2 into the following ratios: $1 - p$ in the upstream half, and $p < 1 - p$ in the downstream half.

If the construction were stopped at this stage, the last step of the first stage would be to represent the resulting relief by six triangles. But if the construction it to continue, the last step is to draw the three smaller drainage basins shown on Fig.A.11c.

The second stage of construction. Each or the basins at the end of stage one has one "mouth" vertex and five "source" vertices, and otherwise each satisfies

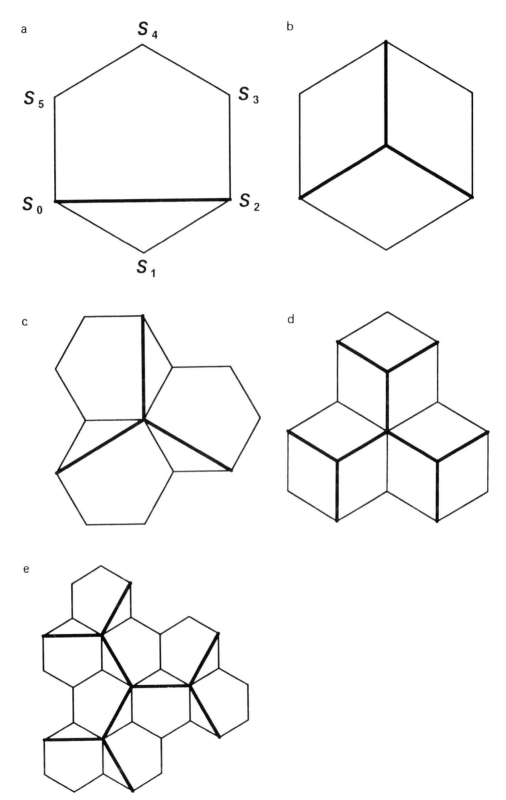

Fig. A.11: Hexagonal tiles with rivers.

the conditions demanded of the initiator. Therefore, one might think that the construction can proceed as in the first stage. However, a major complication arises: some of the new hexagon's source vertices are shared between two neighboring hexagons. This injects context dependence, and demands a new rule to break the deadlock. Between several possibilities, the rule we have adopted is this: when two interpolators of altitude are in conflict, choose the higher.

Exploratory implementations of the above novel algorithms by F.Kenneth Musgrave and Réjean Gagné are stunning, and suggest that the procedures advanced in this text may revive the topic of fractal landscapes.

Appendix B

An eye for fractals

Michael McGuire

My point of view on fractals as a photographer is an inversion of that of the computer graphicists. Their implicit goal in the images they synthesize is photorealism. The inclusion of fractals was a big step toward that goal. It is a piece of what makes reality "real". So I went looking for fractals with my camera. It was knowing they were there that made them visible. By analogy, in 1878 Eadweard Muybridge made the first flash photographs of a galloping horse, proving that at a point in its stride all of its hoofs were off the ground. This settled a bet and from then on painters painted all hoofs off the ground at once. As photography informed painting, so computer graphics informs photography.

Photography is most of all an art of recognition and composition. Repetition has its place as an element of composition. The fractal concept is a major and transcending enrichment of that element. There is an enhancing tension between the intricacy of fractals and that canon of composition which calls for simplicity or unity or one subject at a time. The tension is resolved yet remains due to the unifying power of the fractal concept. That is, the intricacy arises from the recursive application of a simple underlying law. A composition comes together not from making an explicit statement of what is the particular law or algorithm for a fractal seen, but from an intuitive recognition that this is what is there.

Artists have always been keen observers of the natural world, and one would expect to find fractals in their work if one looked. Leonardo's sketches of turbulent water display this. I can think of no more explicitly fractal image than Hokusai's *Great Wave of Kanazawa*. The great capital C shape of the wave on

261

the left as it threatens to engulf the small sampan like boat repeats itself in ever smaller waves and wavelets throughout the picture. It is there in contemporary landscape photography. I have noticed it especially in works by Brett Weston, Don Worth, and William Garnett. In his Basic Photo Series book *Natural Light Photography*, first published in 1952, Ansel Adams wrote about photographing textures in a way that demonstrated a deep and implicit understanding of the fractal concept, lacking perhaps only label we now use. He surely had in mind his *Mt. Williamson from Manzanar* picture with its large boulder in the foreground and field of boulders behind it, and the echoing shapes of the mountains behind them, with the late afternoon sunlight streaming through the clouds. He wrote,

> *Consider photographing rocky landscapes, another instance that presents serious problems in textural rendition. Beyond a certain distance, a great field of granite boulders will appear as perfectly smooth stones, the natural textures being beyond the resolving power of the lens and/or the emulsion. In order to suggest the substance of these stones it is necessary to include in the very near foreground a boulder in which the texture is adequately revealed. You can then say that the photograph "reads well". While you cannot see the texture in the distant boulders, you can see it in the near boulder, and you assume that all the boulders are the same material. It is this* awareness of substance *that is of vital importance in interpretative photography. The photograph no matter what its function may be, must "read" clearly.*

Adams is telling us that fractals are essential.

It is worth recalling a tradition of mathematics and geometry influencing and affecting photography. In particular a book *The Curves of Life* was published in 1914 by Theodore Andrea Cook. Its emphasis was on spiral growth patterns. Its influence on Edward Steichen in his photographs of sunflowers and on Edward Weston in his nautilus shells has been well documented. The photographs accompanying this essay are an attempt to carry on this tradition. They were made with an awareness of the fractal idea and an intention to display it. As to technical details, they were done in large format which favors the capture of detail over the widest range of scales — fractals.

Fig. B.0: Mount Williamson, Sierra Nevada. Photograph by Ansel Adams, 1944. Courtesy of the trustees of the Ansel Adams Publishing Rights Trust. All rights reserved.

Fig. B.1: California oak tree I, Arastradero Preserve, Palo Alto

Fig. B.2: California oak tree II, Arastradero Preserve, Palo Alto

Fig. B.3: Pahoehoe lava I, Hawaii

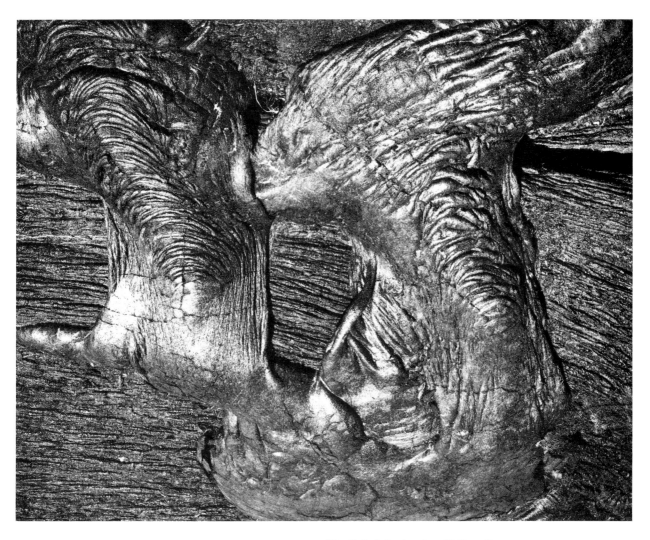

Fig. B.4: Pahoehoe lava II, Hawaii

Fig. B.5: Ohaiu Forest, Kauai, Hawaii

Fig. B.6: Pandanus, Kauai, Hawaii

Fig. B.7: untitled

Fig. B.8: untitled

Appendix C

A unified approach to fractal curves and plants

Dietmar Saupe

C.1 String rewriting systems

In this appendix we introduce so called L-systems or string rewriting systems, which produce character strings to be interpreted as curves and pictures. The method, which is very interesting in itself, fills two gaps that the main chapters leave open. First, rewriting systems provide an elegant way to generate the classic fractal curves, e.g. the von Koch snowflake curve and the space filling curves of Peano and Hilbert. Secondly, the method had been designed by its inventors to model the topology and geometry of plants, in particular of the branching patterns of trees and bushes. Thus, this appendix provides a modeling technique which readily yields relatively realistic looking plants.

In the course of the algorithm a long string of characters is generated. The characters are letters of the alphabet or special characters such as '+', '−', '[', etc. Such a string corresponds to a picture. The correspondence is established via a LOGO-like turtle which interprets the characters sequentially as basic commands such as "move forward", "turn left", "turn right", etc. The main ingredient of the method is the algorithm for the string generation, of course. A first string consisting of only a few characters must be given. It is called the *axiom*. Then each character of the axiom is replaced by a string taken from a table of *production rules.* This substitution procedure is repeated a prescribed number of times to produce the end result. In summary we have that the final picture is completely determined by

- the axiom,

- the production rules,

- the number of cycles,

and the definition for the actions of the turtle, which draws the picture from the output string.

ALGORITHM **Plot-0L-System** (maxlevel)		
Title	Fractal curves and plants using 0L-systems	
Arguments	maxlevel	number of cycles in string generation
Globals	axiom	string containing the axiom of 0L-system
	Kar[]	character array (size num), inputs of production rules
	Rule[]	string array (size num), outputs of production rules
	num	number of production rules
	TurtleDirN	number of possible directions of turtle
Variables	str	string to be generated and interpreted
	xmin, xmax	range covered by curve in x direction
	ymin, ymax	range covered by curve in y direction
BEGIN		
GenerateString (maxlevel, str)		
CleanUpString (str)		
GetCurveSize (str, xmin, xmax, ymin, ymax)		
TurtleInterpretation (str, xmin, xmax, ymin, ymax)		
END		

We will see that in order to specify a complex curve or tree, only a few production rules will suffice. The axiom along with the production rules may be regarded as the genes which control the "growth" of the object. This information can be very small as compared to the complexity of the resulting picture. The same is true for the iterated function systems of Chapter 5. The challenge for iterated functions systems as well as for L-systems is to establish the system information necessary to produce a given object or an object with given properties. This is a topic of current research.

L-systems were introduced by A. Lindenmayer in 1968 for the purpose of modeling the growth of living organisms, in particular the branching patterns of plants. A. R. Smith in 1984 [96] and P. Prusinkiewicz in 1986 [88] incorporated L-systems into computer graphics. This Appendix is based on the work of P.

```
ALGORITHM GenerateString (maxlevel, str)
Title            String generation in 0L-System

Arguments    maxlevel     number of cycles in string generation
             str          output string
Globals      axiom        string containing the axiom of 0L-system
             Kar[]        character array (size num), inputs of production rules
             Rule[]       string array (size num), outputs of production rules
             num          number of production rules
Variables    level        integer, number of string generation cycle
             command      character
             str0         string
             i, k         integer
Functions    strlen(s)    returns the length of string s
             strapp(s, t) returns string s appended by string t
             getchar(s, k) returns k-th character of string s
             strcpy(s, t) procedure to copy string t into string s

BEGIN
    strcpy (str0, axiom)
    strcpy (str, "")
    FOR level=1 TO maxlevel DO
        FOR k=1 TO strlen (str0) DO
            command := getchar (str0, k)
            i := 0
            WHILE (i < num AND NOT command = Kar[i]) DO
                i := i + 1
            END WHILE
            IF (command = Kar[i]) THEN
                str := strapp (str, Rule[i])
            END IF
        END FOR
        strcpy (str0, str)
    END FOR
END
```

Prusinkiewicz. All definitions and most examples are taken from his paper[1], where the interested reader will also find a more detailed account of the history of L-systems and further references.

C.2 The von Koch snowflake curve revisited

The principle of L-systems can be exemplified with the von Koch snowflake curve of Section 1.1.1 (compare Figure 1.3). Let us assume that the LOGO-like turtle is equipped with the following character instruction set:

---

[1] See also his recent publication "Applications of L-systems to computer imagery", to appear in: "Graph Grammars and their Application to Computer Science; Third International Workshop", H. Ehrig, M. Nagl, A. Rosenfeld and G. Rozenberg (eds.), (Springer-Verlag, New York, 1988).

Fig. C.1: The first six stages in the generation of the von Koch snowflake curve. The 0L-system is given by the axiom "F", the angle $\delta = \frac{\pi}{3}$ and the production rule F $\rightarrow$ F−F++F−F.

```
ALGORITHM CleanUpString (str)
Title          Clean out all redundant state preserving commands

Arguments   str          string
Variables   str0         string, initialized to ""
            i            integer
            c            character
Functions   strlen(s)    returns the length of string s
            strappc(s, c) returns string s appended by character c
            getchar(s, k) returns k-th character of string s
            strcpy(s, t) procedure to copy string t into string s

BEGIN
    FOR i=1 TO strlen (str) DO
        c := getchar (str, i)
        IF (c='F' OR c='f' OR c='+' OR c='-' OR c='|'
            OR c='[' OR c=']') THEN
            str0 := strappc (str0, c)
        END IF
    END FOR
    strcpy (str, str0)
END
```

'F' draw a line forward,

'+' turn right by 60°,

'−' turn left by 60°.

We start out with a straight line, denoted by "F". This is the axiom of the von Koch snowflake curve. In stage 1 the line is replaced by a line forward, a left turn, a line, two right turns for a total of 120°, a line, a left turn and another line. In the turtle language this can be written as the string "F−F++F−F". We can ignore the specification of the lengths of the line segments at this point. Subsequently, each line; symbolized by the character 'F', again has to be replaced by the string "F−F++F−F". Thus, in stage 2 we have the string

"F−F++F−F−F−F++F−F++F−F++F−F−F−F++F−F",

and in stage 3 we obtain

"F−F++F−F − F−F++F−F ++ F−F++F−F − F−F++F−F −
F−F++F−F − F−F++F−F ++ F−F++F−F − F−F++F−F ++
F−F++F−F − F−F++F−F ++ F−F++F−F − F−F++F−F −
F−F++F−F − F−F++F−F ++ F−F++F−F − F−F++F−F",

and so forth.

Fig. C.2: The first six stages in the generation of the space filling Hilbert curve. The 0L-system is given by the axiom "X", the angle $\delta = \frac{\pi}{2}$ and the production rules X $\rightarrow$ $-$YF+XFX+FY$-$, Y $\rightarrow$ +XF$-$YFY$-$FX+.

```
ALGORITHM GetCurveSize  (str, xmin, xmax, ymin, ymax)
Title           Compute the size of curve given by string
```

Arguments	str	string
	xmin, xmax	range covered by curve in x direction
	ymin, ymax	range covered by curve in y direction
Variables	i	integer
	command	character
Globals	TurtleX	real x position of turtle
	TurtleY	real y position of turtle
	TurtleDir	direction of turtle (coded as integer)
	TurtleDirN	number of possible directions of turtle
	CO[]	array of TurtleDirN cosine values
	SI[]	array of TurtleDirN sine values
Functions	strlen(s)	returns the length of string s
	getchar(s, k)	returns k-th character of string s

```
BEGIN
    FOR i=0 TO TurtleDirN-1 DO
        CO[i] := cos (2 * Pi * i / TurtleDirN)
        SI[i] := sin (2 * Pi * i / TurtleDirN)
    END FOR
    TurtleDir := 0
    TurtleX := TurtleY := 0.0
    xmax := xmin := ymax := ymin := 0.0
    FOR i=1 TO strlen (str) DO
        command := getchar (str, i)
        UpdateTurtleState (command)
        IF (command='F' OR command='f') THEN
            xmax := MAX (TurtleX, xmax)
            xmin := MIN (TurtleX, xmin)
            ymax := MAX (TurtleY, ymax)
            ymin := MIN (TurtleY, ymin)
        END IF
    END FOR
END
```

In summary we have that when proceeding from one stage to the next we must replace a character 'F' by the string "F−F++F−F", while the characters '+' and '−' are preserved. Thus, the L-system consists of the axiom "F" and the production rules

$$F \rightarrow F{-}F{+}{+}F{-}F, \quad + \rightarrow +, \quad - \rightarrow -.$$

C.3   Formal definitions and implementation

There are different kinds of L-systems. Here we consider only 0L-systems, in which characters are replaced by strings (in pL-systems more complicated substitution rules may apply).

ALGORITHM **UpdateTurtleState** (command)
Title   Change the state of turtle according to given command

Arguments	command	character command
Globals	TurtleX	real x position of turtle
	TurtleY	real y position of turtle
	TurtleDir	direction of turtle (initialized to 0)
	TurtleDirN	number of possible directions (coded as even integer)
	TStackX[]	stack of turtle x positions
	TStackY[]	stack of turtle y positions
	TStackDir[]	stack of turtle directions
	TStackSize	size of turtle stack
	TStackMax	maximal size of turtle stack
	CO[]	array of TurtleDirN cosine values
	SI[]	array of TurtleDirN sine values

```
BEGIN
    IF (command = 'F' OR command = 'f') THEN
        TurtleX := TurtleX + CO[TurtleDir]
        TurtleY := TurtleY + SI[TurtleDir]
    ELSE IF (command = '+') THEN
        TurtleDir := TurtleDir - 1
        IF (TurtleDir < 0) THEN
            TurtleDir := TurtleDirN - 1
        END IF
    ELSE IF (command = '-') THEN
        TurtleDir := TurtleDir + 1
        IF (TurtleDir = TurtleDirN) THEN
            TurtleDir := 0
        END IF
    ELSE IF (command = '|') THEN
        TurtleDir := TurtleDir + TurtleDirN / 2
        IF (TurtleDir > TurtleDirN) THEN
            TurtleDir := TurtleDir - TurtleDirN
        END IF
    ELSE IF (command = '[') THEN
        IF (TStackSize == TStackMax) THEN
            PRINT ("ERROR : Maximal stack size exceeded.")
            EXIT PROGRAM
        END IF
        TStackX[TStackSize] := TurtleX
        TStackY[TStackSize] := TurtleY
        TStackDir[TStackSize] := TurtleDir
        TStackSize := TStackSize + 1
    ELSE IF (command = ']') THEN
        IF (TStackSize == 0) THEN
            PRINT ("ERROR : Stack empty.")
            EXIT PROGRAM
        END IF
        TStackSize := TStackSize - 1
        TurtleX := TStackX[TStackSize]
        TurtleY := TStackY[TStackSize]
        TurtleDir := TStackDir[TStackSize]
    END IF
END
```

```
ALGORITHM TurtleInterpretation (str, xmin, xmax, ymin, ymax)
Title          Plot the curve given by string

Arguments    str          string
             xmin, xmax   range covered by curve in x direction
             ymin, ymax   range covered by curve in y direction
Variables    i            integer
             command      character
             Factor       real
             ix, iy       integers
Globals      xsize, ysize size of screen in pixels
Functions    strlen(s)    returns the length of string s
             getchar(s, k) returns k-th character of string s
             MoveTo ()    move the graphics position
             DrawTo ()    draw a line to the new graphics position
             XScreen (x) = INT (Factor * (x - xmin))
             YScreen (y) = INT (Factor * (y - ymin))

BEGIN
    Factor := MIN ((xsize-1)/(xmax-xmin), (ysize-1)/(ymax-ymin))
    TurtleDir := 0
    TurtleX := TurtleY := 0.0
    MoveTo (XScreen (TurtleX), YScreen (TurtleY))
    FOR i=1 TO strlen (str) DO
        command := getchar (str, i)
        UpdateTurtleState (command)
        IF (command='F') THEN
            DrawTo (XScreen (TurtleX), YScreen (TurtleY))
        ELSE IF (command='f') THEN
            MoveTo (XScreen (TurtleX), YScreen (TurtleY))
        END IF
    END FOR
END
```

Let V denote an alphabet and V^* the set of all words over V. A 0L-system is a triplet $< V, \omega, P >$, where V is the alphabet, $\omega \in V^*$ a nonempty word called the axiom and $P \subset V \times V^*$ is a finite set of production rules. If a pair (c, s) is a production, we write $c \rightarrow s$. For each letter $c \in V$ there is at least one word $s \in V^*$ such that $c \rightarrow s$. A 0L-system is deterministic, if and only if for each $c \in V$ there is exactly one $s \in V^*$ such that $c \rightarrow s$.

In all of the examples we do not specify the alphabet V explicitly, it consists of all characters that occur in the axiom and the production rules. Also, if a specific rule for a character is not stated, then it is assumed to be the identity, i.e. $+ \rightarrow +$, $- \rightarrow -$, and usually $F \rightarrow F$.

The included pseudo code implements the string generation procedure and the graphical interpretation. The program first expands the axiom, then expands the resulting string, and so forth. *CleanUpString()* is called next. It removes all

Fig. C.3: The 8-th stage in the generation of the classic Sierpinsky gasket. The 0L-system is given by the axiom "FXF−−FF−−FF", the angle $\delta = \frac{\pi}{3}$ and the production rules X → −−FXF++FXF++FXF−− and F → FF.

Graphical interpretation of a string : The algorithm which interprets the output string of an L-system (the "turtle") acts upon a character command as follows. We define the state of the turtle as a vector of three numbers denoting the position of the turtle (x-position and y-position) and the direction in which the turtle is heading (an angle). The following character commands are recognized by the turtle :

'F':   move one step forward in the present direction and draw the line,
'f':   move one step forward in the present direction but do not draw the line,
'+':   turn right by an angle given a priori,
'−':   turn left by an angle given a priori,
'|':   turn back (turn by 180^0),
'[':   save the state of the turtle on a stack,
']':   put the turtle into the state on the top of the stack and remove that item from the stack.

All other commands are ignored by the turtle, i.e. the turtle preserves its state.

---

those letters from the string which have significance only for the string reproduction but not for the graphical interpretation. The purpose of this is merely to avoid unnecessary operations. The algorithms *GetCurveSize()* and *TurtleIn-*

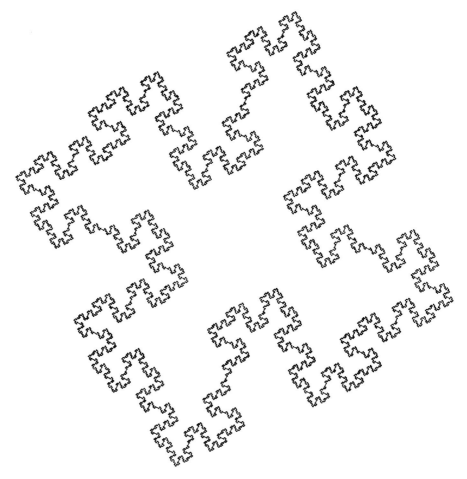

Fig. C.4: The quadratic Koch island (5-th stage) [68]. The 0L-system is given by the axiom "F+F+F+F", the angle $\delta = \frac{\pi}{2}$ and the production rule F $\rightarrow$ F+F−F−−FFF+F+F−F.

terpretation() finally determine the size of the picture and produce the output plot. It should be noted that in the case of a 0L-system an equivalent recursive routine can be written. It would expand one character at a time, according to the corresponding production rule and then recurse for each of the characters of the resulting string until the desired level of detail is achieved. The final characters can immediately be interpreted graphically provided that the overall dimensions of the picture are known a priori.

The routines for string manipulation provided by programming languages differ greatly from one language to another. For our implementation we have borrowed some routines from the C language. However, these should easily be imitated in different languages such as Pascal or Basic.

Fig. C.5: On the left the space filling Peano curve (third stage). The 0L-system is given by the axiom "X", the angle $\delta = \frac{\pi}{2}$ and the production rules X → XFYFX+F+YFXFY−F−XFYFX, Y → YFXFY−F−XFYFX+F+YFXFY. On the right a square Sierpinsky curve (5-th stage). The 0L-system is given by the axiom "F+F+F+F", the angle $\delta = \frac{\pi}{2}$ and the production rule F → FF+F+F+F+FF.

Fig. C.6: Dragon curve (16-th stage). The 0L-system is given by the axiom "X", the angle $\delta = \frac{\pi}{2}$ and the production rules X → X+YF+, Y → −FX−Y.

The pseudo code is not optimized for speed. Many improvements can be made. For example, characters can be identified with their integer ASCII representation. For each of the 128 characters there can be one (possibly empty) rule stored in an array of rules. In this setup the search for the appropriate rule

Fig. C.7: Two bushes generated by 0L-systems. The left one (6-th stage) is given by the axiom "F", the angle $\delta = \frac{\pi}{7}$ and the production rule F → F[+F]F[−F]F. The right one (8-th stage) is given by the axiom "G", the angle $\delta = \frac{\pi}{7}$ and the production rules G → GFX[+G][−G], X → X[−FFF][+FFF]FX. **Figure on page 272** : Bush (5-th stage) with axiom "F", the angle $\delta = \frac{\pi}{8}$ and the production rule F → FF+[+F−F−F]−[−F+F+F].

to apply in the expansion of a letter is simpler and faster. As a second remark let us point out that the system routine *strapp()*, which appends a string to another string, is called very often in the algorithm *GenerateString()*. This process generates very long strings, and as a consequence the performance of *strapp()* may decrease dramatically. It is easy to circumvent this problem, but this again depends to a large extent on the choice of the programming language. Thus, we omit these details here.

Our approach to L-systems can be extended in many ways. We give a few ideas (see [88]).

Fig. C.8: Bush (11-th stage) with axiom "SLFFF", the angle $\delta = \frac{\pi}{10}$ and the production rules S $\rightarrow$ [+++G][− − −G]TS, G $\rightarrow$ +H[−G]L, H $\rightarrow$ −G[+H]L, T $\rightarrow$ TL, L $\rightarrow$ [−FFF][+FFF]F.

1. The strategy for the expansion of characters can be modified so that one or another rule may apply according to preceding and succeeding letters. This is a case of context dependence, and the resulting systems are called pL-systems (pseudo L-systems).

2. All objects generated by deterministic L-systems are fixed. There are no variations. In order to obtain several different specimen of the same species, some randomness must be introduced along with a non-deterministic L-system. Thus there are several possible rules to apply in a given situation. Some probabilities should be attached to these rules a priori. Then a random number generator may decide which of the rules is eventually applied. This procedure is very reminiscent of the iterated function systems in Chapter 5.

3. The turtle which interprets the expanded strings may be allowed to learn a wider vocabulary of commands. E. g. parentheses can be used to group drawing commands which define the boundary of a polygon to be filled. Of course, linestyle and color are parameters of interest. Also, the turtle may be instructed to move and draw in three dimensions. Curved surfaces may be considered in addition to polygons.

Appendix D

Exploring the Mandelbrot set

Yuval Fisher

This appendix describes distance bounds and an algorithm which can be used to generate black and white images of the Mandelbrot set M very quickly. Using these bounds, the algorithm computes disks in $\mathbf{C}$, the complex plane, which are completely inside or outside of the set M. After a relatively small number of these disks are computed and filled, the intricate boundary of the Mandelbrot set emerges. A rough outline of the set is generated almost instantaneously as compared with scanning methods to image M, making the algorithm ideal for real-time exploration of M. The algorithm is based on the ideas of J. Milnor and W. Thurston (see Section 4.2.5) which were expanded by A. Wilkes and J. Hubbard.

The upper bound in Section 4.2.5 on the distance from a point c outside of the Mandelbrot set to the set was used to draw some of the beautiful black and white images of the Mandelbrot set's boundary in this book. If a lower bound on this distance was known, a disk could be drawn around c which would lie entirely outside of the set, and this has the enormous advantage that no other points in that disk need to be checked for exclusion from M. In fact, one such bound on the distance $d(c, M)$ between a point c lying outside of M and M itself is

$$\frac{\sinh G(c)}{2\, e^{G(c)} |G'(c)|} < d(c, M) < \frac{2\, \sinh G(c)}{|G'(c)|} \qquad (\text{D}.1)$$

where $G(c)$ is the potential of the point c (see Eqn. (4.21)).

In the next section, we demonstrate Eqn. (D.1) and then describe an algorithm (see *DBM/M*) that uses this estimate to draw M. In the second part of this appendix, we derive a similar estimate for a point c inside the Mandelbrot set. That is, given a point $c \in M$, we compute the radius of a disk around c which is completely contained in M. Finally, we briefly describe an analogous argument which applies to connected Julia sets.

Readers who are mainly interested in the implementation may want to proceed directly to the boxes containing the algorithm and pseudo code.

D.1   Bounding the distance to M

In this section we compute a lower bound on the distance from a point c to M using two beautiful results of complex analysis. The first is a classical theorem due to P. Koebe, and the second is a relatively new result in complex dynamics due to Douady and Hubbard.

The Koebe $\frac{1}{4}$ theorem states that if $f : D \to \mathbf{C}$ is an analytic function from the open unit disk $D = \{z : |z| < 1\}$ into the complex plane, and if $f(z)$ is one-to-one with $f(0) = 0$ and $|f'(0)| = 1$ then[1] $f(D) \supset D(\frac{1}{4}, 0)$. That is, the point zero is at least a distance of $\frac{1}{4}$ from the boundary of $f(D)$. If $f(0) = a$ and $|f'(0)| = b$ then we can still apply the Koebe theorem by translating and scaling $f(z)$ to satisfy the hypothesis of the theorem. In this case, $f(D) \supset D(\frac{b}{4}, a)$. Note that in order to compute the radius of the disk, we need to know b, the magnitude of the derivative of $f(z)$.

We relate this result to the Mandelbrot set using a theorem of Douady and Hubbard[29]. They proved the connectedness of M by constructing an analytic, one-to-one and onto map

$$\Phi : \mathbf{C} \setminus M \to \mathbf{C} \setminus \overline{D}$$

where, $\overline{D} = \{z : |z| \le 1\}$. We cannot apply the Koebe theorem to this map since Φ is not defined on a disk. However, if we write $r(z) = \frac{1}{z}$ then $r \circ \Phi$ maps $\mathbf{C} \setminus M$ to $D \setminus \{0\}$, and the inverse of this function is almost defined on a disk. Unfortunately, $(r \circ \Phi)^{-1}(z)$ is not defined at 0, but we can *move* the point where $(r \circ \Phi)^{-1}(z)$ is not defined in the following way. Choose some $c_0 \in \mathbf{C} \setminus M$ and let

$$d = r(\Phi(c_0)) = \frac{1}{\Phi(c_0)} \in D \setminus \{0\}.$$

---

[1] $D(r, z)$ denotes the disk of radius r centered at z in $\mathbf{C}$.

The Möbius transformation[2]

$$m_d(z) = \frac{z - d}{1 - \bar{d}z}$$

maps d to 0 and 0 to $-d$. So $m_d \circ r \circ \Phi$ maps $\mathbf{C} \setminus M$ to $D \setminus \{-d\}$. If we restrict[3] the map $(m_d \circ r \circ \Phi)^{-1}(z)$ to the disk $D(|d|, 0) \subset D \setminus \{-d\}$, we obtain a map Ψ on D given by

$$\begin{aligned}\Psi : D &\to \mathbf{C} \setminus M. \\ z &\mapsto (m_d \circ r \circ \Phi)^{-1}(|d|z).\end{aligned}$$

Ψ maps $0 \mapsto c_0$ and satisfies the hypothesis of the Koebe theorem. We can now compute

$$|\Psi'(0)| = R = \frac{|\Phi^2(c_0)| - 1}{|\Phi(c_0)||\Phi'(c_0)|}, \tag{D.2}$$

and then $D(\frac{R}{4}, c_0) \subset \Psi(D)$. That is, the distance from c_0 to M is at least $\frac{R}{4}$. Substituting $G(c) = \log(|\Phi(c)|)$ in Eqn. (D.2) we get

$$R = \frac{2 \sinh G(c_0)}{e^{G(c_0)}|G'(c_0)|}, \tag{D.3}$$

proving Eqn. (D.1).

When c is near M, $G(c)$ is close to 0, and we can use Eqn. (4.30) and the estimate $\sinh G(c) \approx G(c)$ to approximate Eqn. (D.3) by

$$R \approx 2 \frac{1}{|z_n|^{\frac{1}{2^n}}} \frac{|z_n|}{|z_n'|} \log |z_n| \tag{D.4}$$

with $z_n = f_{c_0}^n(c_0)$ and $z_n' = \frac{d}{dc} f_{c_0}^n(c_0)$.

When c_0 is far from M, that is when the estimate $\sinh G(c_0) \approx G(c_0)$ is poor, we can approximate R in Eqn. (D.2) using

$$|\Phi(c_0)| \approx |z_n|^{\frac{1}{2^n}} \tag{D.5}$$

$$|\Phi'(c_0)| \approx \frac{1}{2^n}|z_n|^{\frac{1}{2^n}-1}|z_n'|, \tag{D.6}$$

which are a consequence of Hubbard and Douady's construction of the map $\Phi(c)$ [29].

---

[2] if $d = x + iy$ then $\bar{d}$, the complex conjugate of d, is $\bar{d} = x - iy$. The map $m_d(z)$ is analytic and bijective on D, which is important since we need a one-to-one analytic map in order to apply the Koebe theorem.

[3] This means that we are "throwing away" some points in D in our estimate. A more accurate and somewhat cumbersome bound can be derived by considering the inverse of $m_d \circ r \circ \Phi \circ r : \mathbf{C} \setminus r(M) \to D$.

DBM/M - Distance Bound Method for Mandelbrot set : For c outside of M we want to compute the radius of a disk centered at c which will be completely outside of M. We will iterate $z_{n+1} = z_n^2 + c$ until $|z_n| > R_{max}$ for some $R_{max} > 2$. For n large enough, the $1/|z_n|^{\frac{1}{2^n}}$ term in Eqn. (D.4) is very nearly 1. For example, with $R_{max} = 100$ and $n = 20$, 1 and $1/|z_n|^{\frac{1}{2^n}}$ differ by only 0.01 percent. So in practice we simply ignore this term. Since Eqn. (D.4) is very similar to Eqn. (4.30), we estimate the lower bound on the distance from c to M as one fourth of the estimate of the upper bound computed in *MSetDist()*. That is, for $|z_n|$ large,

$$dist(c, M) \approx \frac{|z_n|}{2|z_n'|} \log |z_n| \qquad (D.7)$$

with $z_{n+1} = z_n^2 + c, z_0 = c$ and $z_{n+1}' = 2z_n z_n' + 1, z_0' = 1$.

When c is near M, this estimate works well. When c is far from M (such as for images of the whole set), it is better to estimate $dist(c, M)$ using Eqn. (D.2) combined with Eqn. (D.5) and Eqn. (D.6). It is easier, however, to simply use Eqn. (D.7) and work only with reasonably close images of M, e.g. for $c = x + iy$ with $x \in [-2.0, 0.5]$ and $y \in [-1.25, 1.25]$, where practice shows that the estimate works.

In the algorithm we scan through the pixels in the image until we find a pixel corresponding to a point c outside of M. We then fill a disk of radius $dist(c, M)$ around c with a label indicating that every point in the disk is not in M. Now we call the algorithm recursively at selected points around the boundary of the disk, and in this way, we remove "bites" from $C \setminus M$ which eat away almost all of $C \setminus M$ (see Figures D.1 and D.2).

Since these bites may not hit every pixel in $C \setminus M$, we continue to scan through the image checking for unmarked pixels, which we mark as belonging to M; to $C \setminus M$; or to the boundary of M, by the following criterion. For an unmarked pixel corresponding to the point $c \in C$, mark c in M if the iterates $|f_c^n(0)|$ remain bounded by R_{max}. Otherwise, compute $dist(c, M)$. A program parameter DELTA determines when to set an unmarked pixel as belonging to the boundary of M. We do this when such a pixel at the point c satisfies $dist(c, M) <$ DELTA. Thus, DELTA determines the "thickness" of the boundary in the image. If $dist(c, M) \geq$ DELTA, we mark c as being in $C \setminus M$.

The program parameter RECUR, determines when the recursion stops. That is, if $dist(c, M) <$ RECUR then we do not call the algorithm recursively. Varying RECUR, determines how quickly and how crudely a rough outline of M is generated.

We can modify the algorithm slightly by varying the parameter RECUR during runtime. By doing this, we produce successively refined images of M very quickly. Initially, we fix RECUR at some large value. When we compute a disk with radius smaller than RECUR, we store its center in a list. When no more disks with radius greater than RECUR are found, we reduce RECUR and call the algorithm recursively at several points along the boundary of the disks on the list. Figure D.2 shows how this modification generates successively refined images of M.

Finally, it is possible to use a somewhat different recursive algorithm when, for example, no primitive routines to fill a disk exists. In this algorithm, we break the image into square regions, and compute the lower bound on the distance to M from the center of each region. If the region is wholly outside of M we mark it as such, otherwise, we break it up into four smaller square regions and repeat the algorithm on each of these. When the regions contain only one pixel, we can repeat the procedure above using DELTA to label the pixel as belonging to the boundary of M or not.

Fig. D.1: The disks computed by the *MDisk( )* algorithm are shown around the Mandelbrot set. The order of the recursion is apparent in the tiling of the disks.

ALGORITHM **FastM** ()

Title	Fast image of Mandelbrot set via distance estimate. Scanning part.
Globals	MSet[][] square array of boolean type, size n by n, initialized to 0
	n image resolution in both x- and y-direction
Variables	ix, iy integers which scan through MSet[][]
Comments	This routine scans through the array MSet[][] which will contain the resulting image. It calls *MDisk( )* which draws the appropriate disks in MSet[] and then calls itself recursively in order to draw a rough outline of *M* quickly. *M* is approximated by the elements of MSet[][] set to zero.

```
BEGIN
    FOR iy = 0 TO n-1 DO
        FOR ix = 0 TO n-1 DO
            IF MSet[ix][iy] = 0 THEN
                MDisk(ix,iy)
            END IF
        END FOR
    END FOR
END
```

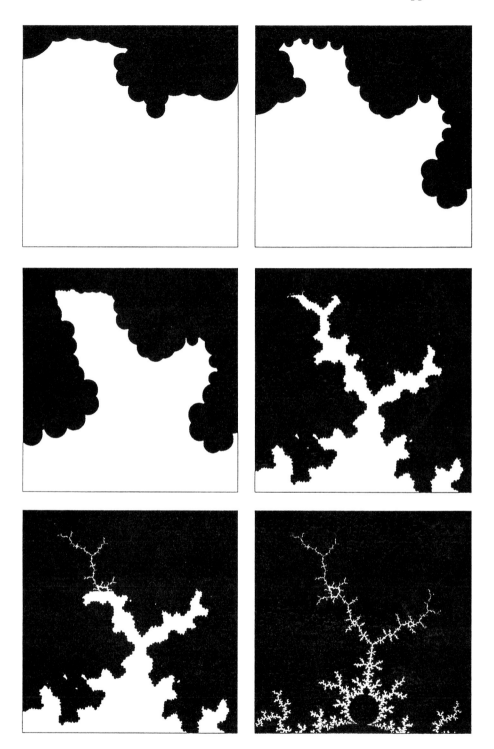

Fig. D.2: Progression of a modified *FastM()* algorithm showing the crude outline of a portion of *M*, successive refinements, and the final scanning stage. The parameter RECUR was first set to 40 and then to 5 in the 696 by 696 pixel images. The images have 20,50,100,1000, 8000 and 60,552 calls to *MDisk()*, respectively. Note that 60,552 is considerably less than the 484,416 points a standard scanning algorithm would need to iterate.

```
ALGORITHM MDisk (ix,iy)
Title           Mark a disk at the point ix, iy in MSet[][]
```

Arguments	ix,iy	point in MSet[][] to be tested
Globals	MSet[][]	square array of boolean type, size n by n, initialized to 0
	n	image resolution in both x- and y-direction
	xmin, ymin	low x-value and y-value of image window
	side	side length of image window - which is square
	maxiter	maximal number of iterations for routine *MSetDist()*
	delta	parameter determining when to mark a pixel as being in the boundary of M
	recur	minimum distance from M for which we call MDisk recursively
Variables	dist	estimate of lower bound on distance to M
	cx,cy	point corresponding to MSet[ix][iy]
	irad	lower bound on distance to M in pixel units
Functions	MSetDist()	returns distance estimate of a point to M
	INT()	returns the integer part of its argument
	FILLDISK()	fill a disk in MSet[][] with the value 1
Comments		delta should be about side / (n-1) . It determines the "thickness" of the boundary.
		recur should be some small integer multiple of delta.
		FILLDISK(Mset[][],ix,iy,irad) should fill a disk centered at Mset[ix][iy] with radius irad.

```
BEGIN
    IF ix>=0 AND ix<n AND iy>=0 AND iy<n AND MSet[ix][iy]=0 THEN
        cx := xmin + side * ix/(n-1)
        cy := ymin + side * iy/(n-1)
        dist := 0.25 * MSetDist(cx,cy,maxiter)
        irad := INT(dist*side)
        IF irad := 1 THEN
            MSet[ix][iy] := 1
        ELSE IF irad > 1 THEN
            FILLDISK(Mset[][],ix,iy,irad)
        ELSE IF dist > delta THEN
            MSet[ix][iy] := 1
        END IF
        IF dist > recur THEN
            IF irad > 1 THEN irad := irad + 1
            MDisk(ix, iy + irad)
            MDisk(ix, iy - irad)
            MDisk(ix + irad, iy)
            MDisk(ix - irad, iy)
            irad := INT(0.5 + irad / sqrt(2))
            MDisk(ix + irad, iy + irad)
            MDisk(ix - irad, iy - irad)
            MDisk(ix - irad, iy + irad)
            MDisk(ix + irad, iy - irad)
        END IF
    END IF
END
```

D.2   Finding disks in the interior of M

In this section we derive a lower bound on the distance from a point $c \in M$ to the boundary of M. As in the previous section, we combine the Koebe $\frac{1}{4}$ theorem with a theorem of Douady and Hubbard.

Douady and Hubbard construct maps from subsets of the interior M to D. The interior of the Mandelbrot set is conjectured to consist of *components* (or regions) where the iterates of $f_c(z) = z^2 + c$ starting at $z = 0$ will converge to a periodic cycle[4] of some period p. A component of the interior of M where the numbers $f_c^n(0)$ converge to a periodic cycle of a fixed period p is called hyperbolic. It is conjectured that all of the components of the interior of M are hyperbolic. For example, the main cardioid of M is a hyperbolic component where the iterates $f_c^n(0)$ converge as $n \to \infty$ to a fixed point (which depends on the point c).

Douady and Hubbard demonstrate that each hyperbolic component of the interior of M can be mapped in a one-to-one fashion onto D. We will apply the Koebe theorem to the inverse of the map they construct, composed with the appropriate Möbius transformation. Specifically, they show that if W is a hyperbolic component of M with $c \in W$, and if $z_0(c), \ldots, z_{p-1}(c)$ is a periodic cycle[5] of period p to which $f_c^n(0)$ converges as $n \to \infty$, then the map

$$\begin{aligned} \rho : W &\to D \\ c &\mapsto \tfrac{d}{dz}(f_c^p)(z_0(c)) \end{aligned}$$

is analytic, one-to-one and onto. If $c_0 \in W$ and $\rho(c_0) = d \in D$ then we can precompose $\rho(c)$ with $m_d(z) = (z-d)/(1-\bar{d}z)$ to get a map $m_d \circ \rho : W \to D$ which maps $c_0 \mapsto 0$. Now the map $(m_d \circ \rho)^{-1}(z)$ satisfies the hypothesis of the Koebe theorem, by construction.

In order to compute the radius for a disk which is contained in W we must find $|\frac{d}{dc}(m_d \circ \rho)^{-1}(0)|$, which we can compute by inverting $|\frac{d}{dc}(m_d \circ \rho)(c_0)|$. Differentiating gives,

$$|\frac{d}{dc}(m_d \circ \rho)(c_0)| = |m_d'(d)||\frac{d}{dc}\frac{d}{dz}(f_c^p)(z_0(c_0))|. \qquad (D.8)$$

Computing the second term is clearer if we write $f_c^p(z) = f^p(c,z)$ so that,

$$\frac{d}{dc}\frac{d}{dz}(f_c^p)(z) = \frac{\partial}{\partial c}\frac{\partial}{\partial z}f^p(c,z) + \frac{\partial}{\partial z}\frac{\partial}{\partial z}f^p(c,z)\frac{dz_0(c)}{dc}. \qquad (D.9)$$

---

[4] A periodic cycle of period p for $f_c(z)$ is a set of numbers $z_0, \ldots, z_{p-1}, z_p = z_0$ satisfying $z_{i+1} = f_c(z_i)$.

[5] The iterates $f_c^n(z)$ and hence the points composing the periodic cycle depend analytically on the parameter c.

This is not the most pleasant expression, but fortunately each term in Eqn. (D.9) can be computed recursively by applying the chain rule to

$$f^{n+1}(c,z) = [f^n(c,z)]^2 + c.$$

This yields,

$$\frac{\partial}{\partial z} f^{n+1} = 2f^n \cdot \frac{\partial}{\partial z} f^n, \tag{D.10}$$

$$\frac{\partial}{\partial c} f^{n+1} = 2f^n \cdot \frac{\partial}{\partial c} f^n + 1, \tag{D.11}$$

$$\frac{\partial}{\partial z} \frac{\partial}{\partial z} f^{n+1} = 2[(\frac{\partial}{\partial z} f^n)^2 + f^n \cdot \frac{\partial}{\partial z} \frac{\partial}{\partial z} f^n], \tag{D.12}$$

$$\frac{\partial}{\partial c} \frac{\partial}{\partial z} f^{n+1} = 2[\frac{\partial}{\partial z} f^n \cdot \frac{\partial}{\partial c} f^n + f^n \cdot \frac{\partial}{\partial c} \frac{\partial}{\partial z} f^n] \tag{D.13}$$

for $n = 0, \ldots, p - 1$ with

$$f^0 = z_0(c_0),$$

$$\frac{\partial}{\partial z} f^0 = 1,$$

$$\frac{\partial}{\partial c} f^0 = \frac{\partial}{\partial z} \frac{\partial}{\partial z} f^0 = \frac{\partial}{\partial c} \frac{\partial}{\partial z} f^0 = 0.$$

Since $z_0(c)$ is periodic, $z_0(c) = f^p(c, z_0(c))$ and we can compute

$$\frac{dz_0(c)}{dc} = \frac{d}{dc}[f^p(c, z_0(c))]$$

$$= \frac{\partial}{\partial c} f^p(c, z_0(c)) + \frac{\partial}{\partial z} f^p(c, z_0(c)) \frac{dz_0(c)}{dc},$$

so,

$$\frac{dz_0(c)}{dc} = \frac{\frac{\partial}{\partial c} f^p(c, z_0(c))}{1 - \frac{\partial}{\partial z} f^p(c, z_0(c))}. \tag{D.14}$$

Combining the equations above we arrive at the desired estimate ,

$$|\frac{d}{dc}(m_d \circ \rho)^{-1}(0)| = R = \frac{A}{B} \tag{D.15}$$

where

$$A = \frac{1}{|m'_d(d)|} = 1 - |\frac{\partial}{\partial z} f^p|^2$$

$$B = |\frac{\partial}{\partial c} \frac{\partial}{\partial z} f^p + \frac{\partial}{\partial z} \frac{\partial}{\partial z} f^p \frac{\frac{\partial}{\partial c} f^p}{1 - \frac{\partial}{\partial z} f^p}|,$$

evaluated at $(c, z) = (c_0, z_0(c_0))$. With this, $D(\frac{R}{4}, c) \subset W \subset M$. That is, all the points which are a distance smaller than $\frac{R}{4}$ from c will be in M also.

DBMI/M - Distance Bound Method for the Interior of the Mandelbrot set : Let c be a point in the interior of M and let $z_0, \ldots, z_{p-1}$ be a periodic cycle of period p to which $f_c^n(0)$ is attracted as $n \to \infty$. We want to compute the radius of a disk centered at c which is contained completely in M. This involves using Eqn. (D.10) to Eqn. (D.14) to recursively compute Eqn. (D.9) and Eqn. (D.15). To overcome the cumbersome complex arithmetic involved, a small complex arithmetic library can be written.

In order to compute the equations recursively, we need to find the period of the periodic cycle to which $f_c^n(0)$ is attracted and a way to estimate z_0. The problem of finding the period does not have a satisfactory solution, and on occasion bad estimates of the lower bound may occur due to an error in computing the period.

The period and z_0 can be found in the following way. As in *MSetDist()*, we iterate $f_c^n(0)$ until n = MAXITER, or until the iterates obtain a large magnitude. In the first case, we assume that $c \in M$ and hope that $f_c^n(0)$ has been attracted close to the periodic cycle. We then check if $f_c^{n+p}(0)$ is near $f_c^n(0)$ for p increasing from 1 to some maximum value. When $f_c^{n+p}(0)$ and $f_c^n(0)$ are near, we guess that p is the period and let $z_0 = f_c^n(0)$. In practice, it is sufficient to check if $|f_c^{n+p}(0) - f_c^n(0)|$ is smaller than 5×10^{-6} times the side length of the region in C we are imaging.

After using Eqn. (D.15) to compute the radius of a disk at c which is contained in M, we incorporate this computation into the *MSetDist()* algorithm so that it returns a bound on the distance to the boundary of M for points in both the exterior and interior of M. The *FastM()* algorithm will then work for the interior of M as well as the exterior.

D.3   Connected Julia sets

The discussion in Section D.1 applies almost identically when we substitute a connected Julia set J for M. The distance estimate turns out to be

$$\frac{\sinh g(z)}{2\, e^{g(z)} |g'(z)|} < d(z, J) < \frac{2\, \sinh g(z)}{|g'(z)|} \qquad (\text{D}.16)$$

where $g(z)$ is the potential of a point z outside of the Julia set J defined in Eqn. (4.15).

As before, the lower bound on the distance is well estimated by

$$\frac{1}{2} \frac{1}{|z_n|^{\frac{1}{2^n}}} \frac{|z_n|}{|z_n'|} \log |z_n|$$

with $z_n = z_{n-1}^2 + c, z_0 = z$; $z_n' = 2 z_{n-1} z_{n-1}', z_0' = 1$; and where c is the complex parameter defining the Julia set for $f_c(z)$. Thus, the only modifications that are required to apply the *DBM/M* algorithm to a connected Julia set involve the recursive equations for z_n and z_n'.

Bibliography

[1] Amburn, P. , Grant, E. , Whitted, T.
Managing geometric complexity with enhanced procedural methods
Computer Graphics 20,4 (1986)

[2] Aono, M. , and Kunii, T.L.
Botanical tree image generation
IEEE Computer Graphics and Applications 4,5 (1984) 10–33

[3] Barnsley, M.F. and Demko, S.
Iterated function systems and the global construction of fractals
The Proceedings of the Royal Society of London A399 (1985) 243–275

[4] Barnsley, M.F. , Ervin, V. , Hardin, D.and Lancaster, J.
Solution of an inverse problem for fractals and other sets
Proceedings of the National Academy of Sciences 83 (1985)

[5] Barnsley, M.F.
Fractal functions and interpolation
Constructive Approximation 2 (1986) 303–329

[6] Barnsley, M.F. , Elton, J.
A new class of Markov processes for image encoding
To appear in : Journal of Applied Probability

[7] Barnsley, M.F.
Fractals Everywhere
to appear, (Academic Press, 1988)

[8] Beaumont, G.P.
Probability and Random Variables
(EllisHorwood Limited, 1986)

[9] Becker K.-H. and Dörfler, M.
Computergraphische Experimente mit Pascal
(Vieweg, Braunschweig, 1986)

[10] Bedford, T.J.
Dimension and dynamics for fractal recurrent sets
Journal of the London Mathematical Society 2,33 (1986) 89–100

[11] Berry, M.V. and Lewis, Z.V.
      On the Weierstrass-Mandelbrot fractal function
      Proc. R. Soc. Lond. A370 (1980) 459–484

[12] Binnig, G. and Rohrer, H.
      The scanning tunneling microscope
      Scientific American (August 1985) 50ff.

[13] Blanchard, P.
      Complex analytic dynamics on the Riemann sphere
      Bull. Amer. Math. Soc. 11 (1984) 85–141

[14] Bouville, C.
      Bounding ellipsoids for ray-fractal intersections
      Computer Graphics 19,3 (1985)

[15] La Brecque, M.
      Fractal Applications
      Mosaic 17,4 (1986) 34–48

[16] Brolin, H.
      Invariant sets under iteration of rational functions
      Arkiv f. Mat. 6 (1965) 103–144

[17] Cochran, W. T. et al.
      What is the Fast Fourier Transform ?
      Proc. IEEE 55 (1967) 1664–1677

[18] Coquillart, S. and Gangnet
      Shaded display of digital maps
      IEEE Computer Graphics and Applications 4,7 (1984)

[19] Carpenter, L.
      Computer rendering of fractal curves and surfaces
      Computer Graphics (1980) 109ff.

[20] Clark, K.
      Landscape Painting
      (Charles Scribner's Sons, New York, 1950)

[21] Coulon, F.de
      Signal Theory and Processing
      (Artech House, 1986)

[22] Coxeter, H.S.M.
      The Non-Euclidean symmetry of Escher's picture 'Circle Limit III'
      Leonardo 12 (1979) 19–25

[23] Demko, S. , Hodges, L. , and Naylor, B.
      Construction of fractal objects with iterated function systems
      Computer Graphics 19,3 (July 1985) 271–278

[24] Devaney, R. L.
A piecewise linear model for the zones of instability of an area preserving map.
Physica D10 (1984) 387–393

[25] Devaney, R. L.
An Introduction to Chaotic Dynamical Systems
(Benjamin Cummings Publishing Co. , Menlo Park, 1986)

[26] Devaney, R. L.
Chaotic bursts in nonlinear dynamical systems
Science 235 (1987) 342–345

[27] Dewdney, A.K. ,
Computer Recreations: Exploring the Mandelbrot set,
Computer Graphics 20,4 (1985) 16ff.

[28] Dewdney, A.K. ,
Computer Recreations: Beauty and profundity: the Mandelbrot set and a flock of its cousins called Julia sets
Scientific American (November 1987) 140–144

[29] Douady, A. , Hubbard, J.H.
Iteration des polynomes quadratiques complexes
CRAS Paris 294 (1982) 123–126

[30] Douady, A. and Hubbard, J.H.
On the dynamics of polynomial-like mappings
Ann. Sci. École Normale Sup., %4e série, t18 (1985) 287–343

[31] Diaconis, P. and Shahshahani, M.
Products of random matrices and computer image generation
Contemporary Mathematics 50 (1986) 173–182

[32] Elton, J.
An ergodic theorem for iterated maps
To appear in : Journal of Ergodic Theory and Dynamical Systems

[33] Escher, M.C.
The World of M.C. Escher
(H.N. Abrams, New York, 1971)

[34] Falconer, K.J.
The Geometry of Fractal Sets
(Cambridge University Press, Cambridge, 1985)

[35] Family, F. and Landau, D.P. (editors)
Kinetics of Aggregation and Gelation
(North-Holland, New York, 1984)

[36] Family, F. and Landau, D.P. (editors)
      Kinetics of Aggregation and Gelation
      (North-Holland, New York, 1984)

[37] Fatou, P.
      Sur les équations fonctionelles
      Bull. Soc. Math. Fr. 47 (1919) 161–271, 48 (1920) 33–94, 208–314

[38] Fishman, B. and Schachter, B.
      Computer display of height fields
      Computers and Graphics 5 (1980) 53-60

[39] Foley, J.D. and van Dam, A.
      Fundamentals of Interactive Computer Graphics
      (Addison-Wesley, Reading, Mass., 1982)

[40] Fournier, A. , Fussell, D. and Carpenter, L.
      Computer rendering of stochastic models
      Comm. of the ACM 25 (1982) 371–384

[41] Fournier, A. and Milligan, T.
      Frame buffer algorithms for stochastic models
      IEEE Computer Graphics and Applications 5,10 (1985)

[42] Freeman, J. J.
      Principles of Noise
      (John Wiley & Sons, Inc. , New York, 1958)

[43] Fricke, R. and Klein, F.
      Vorlesungen über die Theorie der automorphen Funktionen
      (Teubner, Leipzig, 1897) and (Johnson, reprint)

[44] Galambos, J.
      Introductory probability theory
      (Marcel Dekker Inc., 1984)

[45] Gardner, M.
      White and brown music, fractal curves, and one-over-f noise
      Scientific American (April 1978) 16ff.

[46] Gonzalez, R.C. and Wintz, P.
      Digital Image Processing
      (Addison-Wesley, Reading, Mass., 1987)

[47] Guckenheimer, J. and McGehee, R.
      A proof of the Mandelbrot N^2 conjecture
      Report of the Mittag-Leffler Institute, Djursholm, Sweden

[48] Hata, M.
      On the structure of self-similar sets
      Japan Journal of Applied Mathematics 2,2 (1985) 381–414

[49] Harrington, S.
Computer Graphics - A Programming Approach
(McGraw Hill, New York, 1987)

[50] Hearn, D. and Baker, M.P.
Computer Graphics
(Prentice-Hall, Englewood Cliffs, N.J., 1986)

[51] Hénon, M.
A two-dimensional mapping with a strange attractor
Commun. Math. Phys. 50 (1976) 69–77

[52] Hentschel, H.G.E. and Procaccia, I.
The infinite number of generalized dimensions of fractals and strange attractors
Physica 8D (1983) 435–444

[53] Hofstadter, D.R.
Strange attractors : Mathematical patterns delicately poised between order and chaos
Scientific American 245 (May 1982) 16–29

[54] Hutchinson, J.
Fractals and self-similarity
Indiana University Journal of Mathematics 30 (1981) 713–747

[55] Julia, G.
Sur l' iteration des fonctions rationnelles
Journal de Math. Pure et Appl. 8 (1918) 47–245

[56] Jürgens, H. , Peitgen, H.-O. and Saupe, D.
The Mandelbrot Set: A Computer Animation of Complex Dynamical Systems
(Institut für den Wissenschaftlichen Film, Göttingen, 1988) (to appear)

[57] Jürgens, H. , Saupe, D. and Voss, R.
Fast rendering of height fields
in preparation

[58] Kajiya, J.T.
New techniques for ray tracing procedurally defined objects
Computer Graphics 17,3 (1983)

[59] Kawaguchi, Y.
A morphological study of the form of nature
Computer Graphics 16,3 (1982)

[60] Lewis, J.P.
Methods for stochastic spectral synthesis
ACM Transactions on Graphics (1987)

[61] Liu, S.H.
Fractals and their applications in condensed matter physics
Solid State Physics 39 (1986) 207ff.

[62] Lovejoy, S. and Mandelbrot, B.B.
Fractal properties of rain, and a fractal model
Tellus 37A (1985) 209–232

[63] Mandelbrot, B.B. and Ness, J.W.van
Fractional Brownian motion, fractional noises and applications
SIAM Review 10,4 (1968) 422–437

[64] Mandelbrot, B. B. and Wallis, J. R.
Some long-run properties of geophysical records
Water Resources Research 5 (1969) 321–340

[65] Mandelbrot, B.B.
Intermittent turbulence in self-similar cascades: Divergence of higher moments and dimension of the carrier
J. Fluid Mech. 62 (1974) 331–358

[66] Mandelbrot, B.B.
Fractals: Form, Chance, and Dimension
(W.H.Freeman and Co., San Francisco, 1977)

[67] Mandelbrot, B.B.
Fractal aspects of the iteration of $z \mapsto \lambda z(1-z)$ for complex λ and z
Annals NY Acad. Sciences 357 (1980) 249–259

[68] Mandelbrot, B.B.
The Fractal Geometry of Nature
(W.H.Freeman and Co., New York, 1982)

[69] Mandelbrot, B.B.
Comment on computer rendering of fractal stochastic models
Comm. of the ACM 25,8 (1982) 581–583

[70] Mandelbrot, B.B.
Fractal curves osculated by sigma-discs, and construction of self-inverse limit sets
Mathematical Intelligencer 5,2 (1983) 9–17

[71] Mandelbrot, B.B.
Fractals in physics: Squig clusters, diffusions, fractal measures, and the unicity of fractal dimensionality
J. Stat. Phys. 34 (1984) 895–930

[72] Mandelbrot, B.B.
Self-affine fractals and the fractal dimension
Physica Scripta 32 (1985) 257–260

Bibliography   303

[73] Mandelbrot, B.B., Gefen, Y., Aharony, A. and Peyriére, J.
Fractals, their transfer matrices and their eigen-dimensional sequences
J. Phys. A 18 (1985) 335–354

[74] Mastin, G.A. , Watterberg, P.A. and Mareda, J.F.
Fourier synthesis of ocean scenes
IEEE Computer Graphics & Applications, (March 1987) 16–23

[75] May, R.M.
Simple mathematical models with very complicated dynamics
Nature 261 (1976) 459–467

[76] Miller, G.S.P.
The definition and rendering of terrain maps
Computer Graphics 20,4 (1986) 39–48

[77] Milnor, J.
Self-similarity and hairiness in the Mandelbrot set
Institute for Advanced Study, preprint

[78] Musha, T. and Higuchi, H.
The 1/f fluctuation of a traffic current on an expressway
Jap. J. Appl. Phys. 15 (1976) 1271–1275

[79] Norton, A.
Generation and display of geometric fractals in 3-D
Computer Graphics 16,3 (1982) 61–67

[80] Oppenheimer, P.E.
Real time design and animation of fractal plants and trees
Computer Graphics 20,4 (1986)

[81] Orbach, R.
Dynamics of fractal networks
Science 231 (1986) 814–819

[82] Peitgen, H.-O. and Richter, P.H.
Die unendliche Reise
Geo 6 (Juni 1984) 100–124

[83] Peitgen, H.-O. and Richter, P.H.
The Beauty of Fractals
(Springer-Verlag, Berlin, 1986)

[84] Peitgen, H.-O. , Saupe, D. and Haeseler, F.v.
Cayley's problem and Julia sets
Math. Intelligencer 6 (1984) 11–20

[85] Pietronero, L. and Tosatti, E. (editors)
Fractals in Physics
(Elsevier Science Publishers B.V., 1986)

[86] Press, W.H. , Flannery, B.P. , Teukolsky, S.A. and Vetterling, W.T.
Numerical Recipes
(Cambridge University Press, Cambridge, 1986)

[87] Priestley, M.B.
Spectral Analysis and Time Series, Vol. 1 Univariate Series
(Academic Press, 1981)

[88] Prusinkiewicz, P.
Graphical applications of L-systems
Proc. of Graphics Interface 1986 – Vision Interface (1986) 247–253

[89] H. K. Reghbati
An overview of data compression techniques
Computer 14,4 (1981) 71–76

[90] Reif, F.
Fundamentals of statistical and thermal physics
(McGraw-Hill, 1965)

[91] Robinson, F.N.H.
Noise and Fluctuations
(Clarendon Press, Oxford, 1974)

[92] Rogers, David F.
Procedural Elements of Computer Graphics
(McGraw-Hill, New York, 1985)

[93] Saupe, D.
Efficient computation of Julia sets and their fractal dimension
Physica 28D (1987) 358–370

[94] Schuster, H.G.
Deterministic Chaos — An Introduction
(Physik Verlag, Weinheim, 1984)

[95] Shlesinger, M.F. , Mandelbrot, B.B. , and Rubin, R.J. (editors)
Proceedings of a Symposium on Fractals in the Physical Sciences
J. Stat. Phys. 36 (1984)

[96] Smith, A.R.
Plants, fractals, and formal languages
Computer Graphics 18,3 (1984) 1–10

[97] Stanley, H.E. and Ostrowsky, N. (editors)
On Growth and Form: Fractal and Non-Fractal Patterns in Physics
(Martinus Nijhoff Publishers, Dordrecht, Netherlands, 1986)

[98] Sullivan,D.
Quasiconformal homeomorphisms and dynamics I
Ann. Math. 122 (1985) 401–418

[99] Sullivan,D.
Quasiconformal homeomorphisms and dynamics II, III
preprints

[100] Voss, R.F. and Clarke, J.
'1/f noise' in music and speech
Nature 258 (1975) 317–318

[101] Voss, R. F. and Clarke, J.
1/f Noise in Music: Music from 1/f Noise
J. Accous. Soc. Am. 63 (1978) 258–263

[102] Voss, R. F.
1/f (flicker) noise: A brief review
Proc. 32rd Annual Symposium on Frequency Control, Atlantic City, (1979) 40–46

[103] Voss, R. F.
Random fractal forgeries
in : Fundamental Algorithms for Computer Graphics,
R.A. Earnshaw (ed.)
(Springer-Verlag, Berlin, 1985) 805–835

[104] Voss, R. F. , Laibowitz, R.B. and Alessandrini, E.I.
Fractal geometry of percolation in thin gold films
in : Scaling Phenomena in Disordered Systems
R. Pynn and A. Skjeltorp (eds.)
(Plenum, New York, 1985) 279–288

[105] Voss, R. F.
Random fractals: Characterization and measurement
Physica Scripta T13 (1986) 27–32

[106] T. A. Welch
A technique for high performance data compression
Computer 17,6 (1984) 8–19

[107] Yaglom, A.
An Introduction to the Theory of Stationary Random Functions
(Prentice-Hall, 1962) and reprint (Dover, 1973)

[108] *Extended Abstracts: Fractal Aspects of Materials*
(Materials Research Society, Pittsburg) (1984), (1985), (1986)

Index

H.-O. Peitgen and P.H. Richter

The Beauty of Fractals

Images of Complex Dynamical Systems

1987 Award
for Distinguished
Technical Communication

Intellectually stimulating and full of beautiful color plates, this book by German scientists Heinz-Otto Peitgen and Peter H. Richter is an unusual and unique attempt to present the field of Complex Dynamics. The astounding results assembled here invite the general public to share in a new mathematical experience: to revel in the charm of fractal frontiers.

In 88 full color pictures (and many more black and white illustrations), the authors present variations on a theme whose repercussions reach far beyond the realm of mathematics. They show how structures of unseen complexity and beauty unfold by the repeated action of simple rules. The implied unpredictability of many details in these processes, in spite of their complete determination by the given rules, reflects a major challenge to the prevailing scientific conception.

- Learn more about the Mandelbrot set and enjoy the beauty and complexity of this fascinating object.
- Experiment with fractals using the algorithms outlined in the book.
- Four invited contributions, by leading scientists—including Benoit Mandelbrot—and one artist, complement the book with further background and provocative views about the relation of science to art.

". . . With its stunning images in black and white and in color, it is both a mathematics textbook and a coffee table adornment . . ."   **Scientific American**

". . . But until very recently not even a science-fiction writer would have anticipated this picture book celebrating the incredibly rich geometry encoded in the humble quadratic equation . . ."   **Nature**

1986/199 pp./184 figures in 221 separate illus., mostly in color/hardcover/$39.00/ISBN 0-387-15851-0

Springer

Springer-Verlag
Berlin Heidelberg New York
London Paris Tokyo